LEARNING AND TEACHING FOR MATHEMATICAL LITERACY

Typically, most people don't realise when and how they can use the mathematics they were taught in high school – yet many of the mathematical ideas and skills can be a powerful tool for understanding how the world works. *Learning and Teaching for Mathematical Literacy* addresses this situation, offering practical strategies for developing a broader vision of mathematical literacy in the classroom and recognising the importance of maintaining these skills into adult life. Linked to the material explored throughout this book, classroom activities and lesson materials are freely available for use via the QR codes included in each chapter.

Filled with case studies and classroom activities, chapters tackle several topics:

- Describing a framework for a broader vision of mathematical literacy – what is it, and why is it important?
- Teaching mathematical literacy in the classroom.
- Applying mathematical literacy to 'real-life' scenarios: My dad is buying a new dishwasher. Should he buy the extended warranty on offer? My phone works fine, but I've been offered an upgrade. How should I decide whether to take it?
- The role of technology in teaching mathematical literacy.
- Designing mathematical measures for real-world quantities.

Firmly grounded by practical applications for the classroom and beyond, this is an essential handbook for any teacher, teaching assistant, or mathematics subject lead who wishes to develop their students' mathematical literacy skills. This is also an ideal resource for those delivering or enrolled in teacher preparation courses.

Hugh Burkhardt, long time Professor of Mathematical Education and Director of Nottingham University's internationally renowned Shell Centre, is an applied mathematician and strategic researcher-designer in education. Working in both the United Kingdom and the United States, he is particularly interested in the challenges of achieving change in school systems.

Daniel Pead has been an IT leader of the Shell Centre team since spending his gap year working on the *Investigations on Teaching with Microcomputers as an Aid* project in 1983. He has worked on the Centre's contributions to *reSolve, Bowland Maths*, and *World Class Tests*, amongst others. He is particularly interested in the design of computer-based material for teaching and assessment – preferably both together!

Kaye Stacey is an internationally renowned researcher, teacher educator, and a prolific designer of educational materials. She is Emeritus Professor Mathematics Education at the University of Melbourne and chaired the Expert Group for the 2012 PISA assessments of Mathematical Literacy.

All three authors have received awards for contributions to educational design in STEM subjects from ISDDE, the International Society for Design and Development in Education.

IMPACT (Interweaving Mathematics Pedagogy and Content for Teaching)

The Learning and Teaching of Algebra
Ideas, Insights and Activities
Abraham Arcavi, Paul Drijvers, Kaye Stacey

The Learning and Teaching of Geometry in Secondary Schools
A Modeling Perspective
Pat Herbst, Taro Fujita, Stefan Halverscheid, Michael Weiss

The Learning and Teaching of Mathematical Modelling
Mogens Niss, Werner Blum

The Learning and Teaching of Number
Paths Less Travelled Through Well-Trodden Terrain
Rina Zazkis, John Mason, Igor' Kontorovich

The Learning and Teaching of Calculus
Ideas, Insights and Activities
John Monaghan, Robert Ely, Márcia M.F. Pinto, Mike Thomas

The Learning and Teaching of Statistics and Probability
A Perspective Rooted in Quantitative Reasoning and Conceptual Coherence
Luis Saldanha, Neil J. Hatfield, Egan J Chernoff, Caterina Primi

For more information about this series, please visit: IMPACT: Interweaving Mathematics Pedagogy and Content for Teaching - Book Series - Routledge & CRC Press

LEARNING AND TEACHING FOR MATHEMATICAL LITERACY

Making Mathematics Useful for Everyone

Hugh Burkhardt, Daniel Pead, and Kaye Stacey

Routledge
Taylor & Francis Group
LONDON AND NEW YORK

Designed cover image: © Shutterstock

First published 2024
by Routledge
4 Park Square, Milton Park, Abingdon, Oxon OX14 4RN

and by Routledge
605 Third Avenue, New York, NY 10158

Routledge is an imprint of the Taylor & Francis Group, an informa business

British Library Cataloguing-in-Publication Data
A catalogue record for this book is available from the British Library

ISBN: 978-1-032-30116-7 (hbk)
ISBN: 978-1-032-30117-4 (pbk)
ISBN: 978-1-003-30350-3 (ebk)

DOI: 10.4324/9781003303503

Typeset in Sabon
by KnowledgeWorks Global Ltd.

CONTENTS

Supplementary material for this book and up-to-date links to the online resources cited can be found online by scanning this QR code or visiting ltml.mathlit.org.

SERIES FOREWORD

IMPACT, an acronym for *Interweaving Mathematics Pedagogy and Content for Teaching*, is a series of textbooks dedicated to mathematics education and suitable for teacher education. The leading principle of the series is the integration of mathematics content with topics from research on mathematics learning and teaching. Elements from the history and the philosophy of mathematics, as well as curricular issues, are integrated as appropriate.

In mathematics, there are many textbook series representing internationally accepted canonical curricula, but such a series has so far been lacking in mathematics education. It is the intention of IMPACT to fill this gap.

Most of the books in the series focus on fundamental conceptual understanding of the central ideas and relationships of different branches of mathematics. These central ideas and relationships serve as organisers for the structure of each book. This book is rather different – more outward-looking. It is built around the extra power that mathematics can give to people in their everyday lives, interweaving the pedagogy and content that teaching for the development of mathematical literacy involves for teacher educators, teachers and their students in school.

Most students spend many hours a week for many years studying mathematics. Historically, it has been the gateway to well-paid jobs – bookkeepers used to need reliable skills in arithmetic while some professions, like engineering, used algebra. But in this technological age, where those skills are rarely needed outside the classroom, why does mathematics now have so much more curriculum time than, say, Music – also an important and beautiful aspect of human culture? What should school Mathematics now offer the majority of students who are not going into STEM-heavy professions. This book offers an answer, both theoretical and practical.

ACKNOWLEDGEMENTS

We would like to thank the Editors of the IMPACT series, Günter Törner, Tommy Dreyfus, Ghislaine Gueudet, and Nathalie Sinclair for their invitation to write this book. They and our families have provided continuing support throughout. As to what you read, while the responsibility is ours, it reflects many decades of working with wonderful colleagues around the world, notably:

In the UK, the team at the Shell Centre for Mathematical Education of the University of Nottingham, including (in alphabetical order) Alan Bell, Barbara Binns, Jon Coupland, Rita Crust, Clare Dawson, Sheila Evans, Colin Foster, Rosemary Fraser, John Gillespie, Steve Maddern, Andy Noyes, Richard Phillips, John Pitts, Jim Ridgway and Geoff Wake, along with other colleagues from a variety of collaborating institutions. David Spiegelhalter, Daniel Burkhardt Cerigo, Laura Downton, Andy Jervis, Joe Fawcett and Leslie Dietiker helped with the chapters where their input is acknowledged.

Special mention must be made of Malcolm Swan, lead designer on so many Shell Centre projects whose products are described here, and with whom one of us (HB) shared the first Emma Castelnuovo Award of the International Commission on Mathematical Instruction for "innovative, influential work in the practice of mathematics education".

In Australia Gary Asp, Terry Beeby, Lucy Bates, Lynda Ball, Jill Brown, Susie Groves, Brian Low, John Malone, Katie Makar, Barry McCrae, Robyn Pierce, Steve Thornton, Beth Price, Vicki Steinle, Gloria Stillman, Ross Turner, David Leigh-Lancaster, Carly Sawatzki, Vern Treilibs, and Jill Vincent.

In the United States, Mary Bouck, Phil Daro, David Foster, Diane Schaefer, Judah Schwartz, Ann Shannon, Sandy Wilcox and, especially, Alan Schoenfeld with whom a 40-year collaboration in project leadership and writing about insights gained has proved so valuable – and enjoyable.

Finally, thanks to the team at Routledge, particularly Bruce Roberts and Lauren Redhead, who have produced this book so efficiently and expeditiously. It has been a pleasure to work with them.

INTRODUCTION

Mathematical literacy is, roughly speaking, the ability to use your mathematics in meaningful ways in a wide range of everyday life situations.

My dad is buying a new dishwasher. Should he buy the extended warranty on offer?
Two friends and I are going on a vacation together this summer. How should we set about planning it?
What is the chance that a teenager like my child will be in a shooting at school? How does it compare to other hazards in their life?

These are typical of the challenges that people face in life. This book is about how mathematics can help people to tackle them – and, more generally, to better understand the world they live in and to make better decisions.

Mathematical literacy is now widely recognised as important for everyone. In a world where technology does much of the technical 'heavy lifting' traditionally associated with mathematics and statistics, learning how to see the essentials of life-related situations involves using mathematics, at whatever level, in new ways. In designing the school curriculum, the usefulness of mathematics is a key justification for the large amount of time that all children spend in Mathematics class.

In this book we look at the challenges and opportunities that this goal presents for school mathematics and for teacher education. We aim to offer direct support for:

- *teacher educators*, working in both initial teacher preparation and professional development contexts;

DOI: 10.4324/9781003303503-1

- *student teachers,* especially those preparing to teach teenagers, and the more experienced *teachers they work with*; and ultimately, through their work,
- *students* in schools and colleges to whom those teachers teach mathematics.

The aim is to be readable by student teachers coming from a wide range of mathematical backgrounds and to enable them to make mathematical literacy part of their teaching. Our treatment of mathematical literacy encompasses other 'literacies' that feature in educational discussions – in particular statistical literacy, data literacy and aspects of computer literacy and the mathematical sides of field-specific literacies such as scientific literacy, health literacy, and financial literacy, not to mention language itself.

Because mathematical literacy, sometimes called 'quantitative literacy' or 'numeracy', is unlike most topics in the IMPACT series in being outside the current practice of many mathematics teachers and most textbooks, this book aims to be both:

- a descriptive and analytic text for reading about mathematical literacy – intellectually sound yet lively, and
- a handbook that helps teachers to plan lessons in this new domain.

The form of support we have devised for these objectives is a combination of description and analysis of the concepts involved together with direct help with the challenges of turning these insights into classroom learning activities. To illuminate the descriptions and minimise misunderstanding, we use a lot of examples – tasks and lessons that reflect practical situations where using mathematics can have a payoff. We address the practical challenges with examples of high-quality learning activities that address mathematical literacy, with online links to the actual lesson materials.

Our focus will be on ways in which mathematics can give people more power over real-world situations in the form of deeper understanding through critical thinking about life-related contexts. We aim to develop a 'big picture' approach: using mathematics to help bring out the essentials of a practical situation, to build insight into the situation, and support good decisions without having to go into detail as a specialist might. We explore, and illustrate through examples, everyday-life situations where mathematics can be useful, noting the aspects of mathematics that play central roles. We discuss ways in which the complementary foci of mathematical literacy and more traditional school mathematics can be used to strengthen both.

Our aim throughout this book is to balance insight into the nature and processes of mathematical literacy with a pragmatic focus on what will help teachers to create and deliver effective and enjoyable mathematical literacy lessons in their classrooms. It is our hope that the reader, whether 'leader' or 'student', will find much that is useful, and even more that is interesting, in the chapters that follow.

Structure of the book

The book is structured in three parts. In the first three chapters we set out our approach to mathematical literacy. There follow six chapters that describe and illustrate ways in which mathematics can inform understanding of specific areas that are important in everyday life. We then return to a deeper look at aspects of mathematical literacy that have been just touched on before.

In the first theoretical chapter we describe and analyse the meaning of mathematical literacy as set out, for example, by PISA. We present a Context-Focused Mathematics Framework and illustrate it with complementary uses of mathematics in modelling real-world situations and in getting meaning out of data. The framework pays serious attention to the real-world context at the centre, working alongside both knowing how to use mathematics and knowing about its use. Additionally, the framework highlights productive disposition to use mathematics and critical thinking at every stage. To complement and enrich this description from a practical classroom perspective, the next chapter analyses a varied range of exemplar tasks – some readers may prefer to start here. The third chapter explores teaching mathematics in a way that enhances mathematical literacy, and implications for the pedagogy and content of school mathematics curricula.

There follow six chapters on areas of practical importance discussed in more detail: risk in everyday life; climate change, chosen for its far-reaching consequences; planning for good things in life; looking behind the spin in commerce and politics; the history and consequences of social equality and inequality, and – the most well-trodden area – the roles of money in everyday life. Our 'big picture' approach aims to bring out the essential features of each context without too much detailed calculation. Mathematical literacy is not about becoming an expert but, rather, being able to better understand what experts do – and to formulate critical questions about their assumptions and the conclusions that follow. For example, we present and analyse a lot of data; we encourage the reader to question the selections we have made and the sources we list – a key aspect of being mathematically literate.

These chapters aim to show how mathematical literacy, as a widely accepted learning goal, can become part of the enacted mathematics curriculum. To this end, we offer two kinds of support that session leaders, at whatever level, may like to use to engage their students in active learning. We suggest 'activities' for small group work on issues that arise in the text. We outline lessons that have worked well in practice, many with links to well-engineered lesson materials that are freely available online.

The underlying messages in each of these six chapters on teaching for mathematical literacy is the same – readers might not want to study all of them. However, the chapters show something of the rich variety that mathematical literacy offers for widening students' understanding of the world outside the classroom. It thus makes mathematics part of the broader agenda of critical thinking across school subjects in both the sciences and humanities.

The later chapters return to general issues of mathematical literacy. While the use of computer technology is assumed throughout the book, we now look in more depth at various specific ways it can enhance the learning of mathematical literacy. We then turn to analyse curiosity– what it is and how we may develop this prerequisite for wanting to think about life-related phenomena from outside the classroom. As such it is an important and sometimes neglected part of the productive disposition that is identified in the Context-Focused Mathematics Framework. We look at the process of designing mathematical 'measures' of quantities of interest, examples of which have arisen in all of the earlier chapters. We then look in more depth at the mathematics of computer science that underpins digital technology from the 'know about' perspective that helps people understand the tools that they use every day.

Finally, we return to reflect on what has been covered and how the teaching of mathematical literacy fits into the roles that a teacher of mathematics plays, in the classroom and beyond.

Links and references

We have included formal references to published books and papers in traditional format – as with all such citations, some of these may be available to read online, others may require journal subscriptions or only be available via academic libraries. In addition, we have included many links to resources that are – at the time of writing – publicly available online. Some provide reference material that may help with lesson ideas and planning, others offer substantial collections of carefully designed teaching material. Rather than print long web addresses, these links are in the form of codes like [1A] which can be quickly looked up on the supporting website at ltml.mathlit.org. A summary of links is included at the end of each chapter.

1

WHAT IS MATHEMATICAL LITERACY?

Mathematical literacy is the ability to use one's mathematical capabilities to better understand the world, its structures and events, and with that understanding to make better-informed decisions. In this chapter we explore the meaning of mathematical literacy from a theoretical perspective, sketching its history, the various ways it is described and defined, and its relationship to the different aspects of school mathematics.

The term 'mathematical literacy' came to worldwide prominence at the turn of the century with the advent of PISA, the widely influential OECD *Programme for International Student Assessment*. PISA explicitly sets out to assess mathematical literacy, scientific literacy, and reading literacy. The PISA 2022 Mathematics Framework [1A] uses the following wording, setting out clearly what is assessed:

> Mathematical literacy is an individual's capacity to reason mathematically and to formulate, employ, and interpret mathematics to solve problems in a variety of real-world contexts. It includes concepts, procedures, facts, and tools to describe, explain, and predict phenomena. It helps individuals know the role that mathematics plays in the world and make the well-founded judgments and decisions needed by constructive, engaged and reflective 21st Century citizens.
>
> *(Organisation for Economic Co-operation and Development, 2013, p. 25)*

We believe this broad description summarises the various ways in which mathematics can be valuable to every citizen in their life beyond school, including in the world of work. This is the focus of mathematical literacy in

DOI: 10.4324/9781003303503-2

this book. Stacey and Turner (2015) discuss the issues involved in greater depth than we have space for here.

Work on the development of mathematical literacy as a component of school curricula has flourished for over half a century, much of it under the broader umbrella of mathematical modelling (see e.g. Burkhardt, 2018; Steen et al., 2007). Other terms have been used with much the same meaning. *Quantitative literacy* is often preferred in the United States (Madison & Steen, 2008; Steen, 1999, 2002), whilst the term *Numeracy* was introduced by the UK Crowther Committee in the 1950s (Crowther Report 15–18, 1959) as "the mathematical equivalent of literacy". The OECD's *Programme for the International Assessment of Adult Competencies* (PIAAC) [1B] sets out to assess numeracy which it defines as "the ability to use, apply, interpret, and communicate mathematical information and ideas".

The overall goal of this chapter is to provide a rich description of mathematical literacy as a basis for the discussion of learning and teaching mathematical literacy in the next two chapters. There will be many examples from various important context areas in the chapters that follow. In Section 1.1 we describe our context-focused mathematics framework that builds on and integrates various theoretical approaches to mathematical literacy; this theoretical framework links the diverse aspects we present in this book. In Section 1.2 we explore the distinction between the 'knowing how' and 'knowing about' aspects of mathematics. In Section 1.3 we compare 'theory-driven' and 'data-driven' approaches – constructing mathematical models of practical situations and getting meaning out of data – and we look at these two complementary perspectives in more depth in Sections 1.4 and 1.5. These two skill sets feature throughout the book. In Section 1.6 we discuss and illustrate the central role that technology plays in mathematical literacy, throughout the book and in detail in Chapters 10 and 13. Section 1.7 draws attention to the nomenclature problem: that *mathematical literacy* and even more often *numeracy* are also used in a different, very narrow sense – as is *literacy* itself.

1.1 A context-focused mathematics framework

There is a wide variety of theoretical structures that describe mathematical literacy – what it is, or should be, and its place in the school curriculum, both within mathematics and in other subjects. The review by Geiger et al. (2015) provides an overview of these theoretical models, describing their varying emphasis on 'modelling with mathematics' or 'getting meaning out of data'. Note that we use mathematics and mathematical literacy broadly, so it includes statistics, data literacy, and some aspects of digital literacy.

Building on these theoretical ideas, and our and others' experience, we have identified five core components of mathematical literacy, shown in Figure 1.1.1.

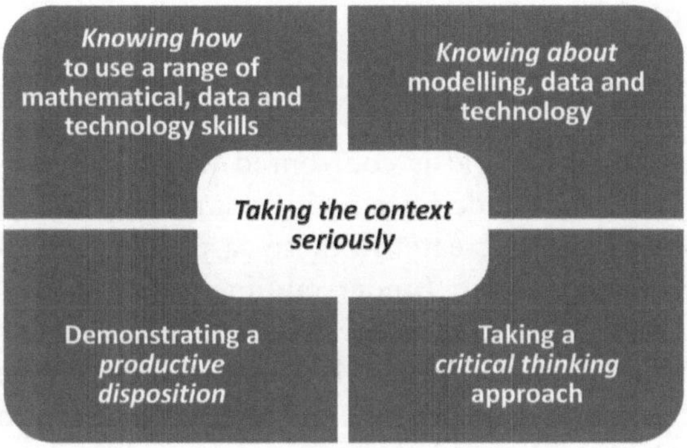

FIGURE 1.1.1 The context-focused mathematics framework.

We assert that mathematical literacy centres around taking the context seriously, knowing about mathematical methods and ideas, and knowing how to use some of them, and also draws on personal cognitive and attitudinal characteristics – notably a critical thinking approach to situations and a disposition towards enquiry.

The five components of mathematical literacy

Taking the context seriously

In mathematical literacy, gaining more understanding of the life-related situation is the priority; in contrast, in school mathematics applications of mathematics are used mainly to exemplify and reinforce mathematical concepts and techniques, which remain the focus. Solving mathematical literacy problems focuses on an understanding of the context that is deeper than in most classroom applications of mathematics.

Knowing how to use a range of mathematical, data, and technology skills

This is obviously central – always bearing in mind that individuals have different levels of, and needs for, mathematical expertise. The word 'skill' here is used broadly to include concepts, processes, and procedures. Importantly, as will become evident, important skills often differ from those most emphasised in traditional school mathematics. They reflect the widespread adoption of information technologies in work and in everyday life, which has had a profound effect on the mathematical knowledge needed for mathematical literacy. We include data to highlight not just the statistical knowledge of how to process it but also aspects of its collection and quality.

Knowing about modelling, data, and technology

Whilst knowing how to create and adapt some mathematical models is at the heart of mathematical literacy, the later chapters demonstrate that a citizen in the modern world is confronted with many life-related contexts where the realistic models are highly complex, and well beyond the capacity of non-experts to create. In these cases, 'knowing about' comes to the fore. 'Knowing about' (understanding) the modelling process is essential to interpret conclusions presented, in the light of the assumptions on which the model is based. Mathematical literacy also requires a general appreciation of good practice in collecting, presenting, and interpreting data and a familiarity with the strengths and limitations of digital technology.

Taking a 'critical thinking' approach

Focusing on the context and the limitations of mathematical models together requires the questioning approach characteristic of critical thinking. This is essential in order to solve life-related problems and to contribute productively to discussing and evaluating the arguments of others. In school, mathematics lessons are an opportunity to engage students in reasoning from evidence to present and defend a case. A critical approach to data and its interpretation is a key part of statistical literacy.

Demonstrating a productive disposition

A productive disposition requires confidence in applying mathematics (for most people in most real-life situations, this mathematics will not be complex), an expectation that using mathematics will be informative, and a willingness to use it. Productive disposition also requires a willingness to persist through obstacles, to be prepared to think carefully and work flexibly, exercising initiative. Curiosity about phenomena is an essential precursor of the kinds of thinking summarised above. Whilst teachers can stimulate such thinking in the classroom, to become mathematically literate citizens, students need to become self-motivated. Mathematically literate individuals see mathematics as a useful tool for exploring situations in which they are actively interested. Fostering this may require encouragement to look beyond what is immediately relevant to students today.

The rest of this chapter develops and illustrates these brief summaries. This framework is an advance organiser, which will help to provide coherence across the diverse contexts that we discuss in the chapters that follow.

1.2 'Know how' versus 'Know about'

We start with an important distinction that is not often explicitly set out. Mathematics is unique among school subjects in that it is almost exclusively a *know how* subject with very little *know about*. In schools, Mathematics is focused very strongly on knowing how to perform procedures and solve problems and the concepts that support this work; it generally includes very little *about* the subject – its history, how concepts and ideas developed – and even less about how mathematics beyond school level contributes to modern life. Mathematical literacy does not always depend on a person knowing how to solve a certain problem in all the detail that school mathematics typically requires. Instead, a mathematically literate person knows enough about the concepts and techniques that experts employ to make well-informed and reasonable decisions based on their advice. For example, a person using an online app to see what the repayments would be for a loan does not need to be able to do all the calculations themselves, but they do need to understand that the app builds in assumptions about future interest rates, and need to know enough about exponential behaviour to appreciate that apparently small deviations in interest rates may alter repayments over time substantially. There are many other examples throughout this book.

In summary, people thinking about challenging life-related problems need to know about the context – the real-world situation of interest – and also know about how mathematics (including statistics) is used by experts to help bring out the essentials of what is going on. The balance between the two will vary with the familiarity and complexity of the context. Because many interesting and important life-related systems are often too complex for non-experts to handle the analysis in detail, we aim to help readers to know about and understand the *principles* of that expert-level analysis so that they can form intelligent questions about the assumptions, methods, and conclusions that people have offered. Even here a wide range of mathematical concepts and skills are involved.

1.3 Complementary approaches – 'theory-driven' and 'data-driven'

There are two complementary strands for approaching an interesting problem situation, which we might call 'theory-driven' and 'data-driven'. Sometimes they are called 'analytic' and 'descriptive'. In a 'theory-driven' approach, you start by thinking about the situation and what quantities (variables) seem important and how they may be related to each other (modelled) and what you would expect (predict); this can then be tested with data and evaluated. In the 'data-driven' approach, you start with some data that looks interesting or is of concern, think what it might mean, and then consider how that phenomenon might have arisen. Often, you use a mixture of these approaches, sometimes starting with ideas about relationships, sometimes with data of interest. For both approaches, you have to decide, at least, which variables seem most important; in practice, your choice of variables is often made for you by the data that is available and catches your attention – though, on thinking the situation through critically, other data may be more informative (see e.g. 'excess deaths' rather than 'COVID deaths' in Chapter 4).

Theory-driven approach – modelling a situation

The theory-driven approach to mathematical literacy focuses on using mathematics to help describe and understand the structure of the problem situation. The process is called mathematical modelling. There are many more-or-less-equivalent descriptions of this process. One summary of the various phases involved is shown in in Figure 1.3.1.

How does this process apply to tackling the *Airplane Turnround* task in Figure 1.3.2?

When a typical class is presented with this task, their first response is to just add up the individual times. That's what you do in a mathematics lesson. But once the teacher gets students to recognise that they must take the context seriously – "Could you find a way to do it in less time?" – students recognise that *this is a different game* and begin to think about the *real-world problem*. Is there a way to do any of these things in parallel to reduce the total turnround time? They then *formulate* a model by using their knowledge of the turnround jobs to decide which must be sequential and which can be done in parallel, going on to *solve* for the time for each parallel track by adding the times of its components (a simple instance of *solve*). They *interpret* that result by noting that the longest parallel track determines the turnround time and, preferably in discussion with others, decide if their assumptions are reasonable, calculations are correct, *and* if the result makes sense in the

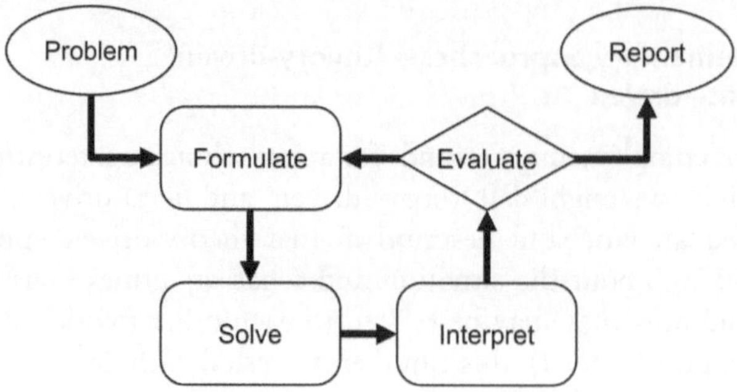

FIGURE 1.3.1 The modelling process.

Source: After Burkhardt, 2008, and, National Governors Association, 2010.

Airplane turnround

Between landing and taking off, the following jobs need to be done on an aircraft.

	Job	Time needed	
A	Get passengers out of the cabin and off the plane	10	minutes
B	Clean the cabin	20	minutes
C	Refuel the plane	40	minutes
D	Unload the baggage from the cargo hold beneath the plane	25	minutes
E	Get new passengers on the plane	25	minutes
F	Load the new baggage into the cargo hold	35	minutes
G	Do a final safety check before take-off	5	minutes

What is the shortest time needed to do all these jobs?

FIGURE 1.3.2 A task demonstrating taking the context seriously.

Source: Adapted from a task in *Bowland Maths* (see Section 6.1).

context – they *evaluate* the model. If it does not seem adequate, they go back to formulate an improved model; if it looks good enough, they *report* the result.

The phases in Figure 1.3.1 just exemplified for this mathematically simple situation, can contain many hidden complexities. Note, in particular, how the thinking has a mainly verbal form in the top half of Figure 1.3.1 and a mathematical form in the lower part, moving back and forth between the two as the modelling process proceeds.

Published variants of the modelling diagram have different emphases and different amounts of detail. The above version emphasises the processes, plunging down from the real world at the top into the mathematics below and re-emerging with potentially useful answers. It makes explicit the iterative nature of most modelling through the 'decision point' symbol *Evaluate* and the important connections to the world of the initial problem through the *Problem* and *Report* bubbles. We shall discuss modelling in more detail in the next section.

As we shall see throughout the book, this theory-driven approach can usefully be applied to many life-related problems; models are rarely *exact* descriptions of the situation but may still give useful insight – hence the aphorism "All models are wrong but some are useful" (Box, 1979, pp. 202–203). Particularly when the situation is complex, properly creating a theory-driven model may be too difficult, yet modelling a few underlying essentials can still be informative. Otherwise, one must turn to empirical data to gain some understanding.

Data-driven approach

The mathematical aspects of 'getting meaning out of data', sometimes called *data literacy*, form the second major strand of this book. Interpreting data presented in tables or graphs of various kinds is a core skill; deciding which quantities are likely to be informative in developing understanding of the situation and then finding that data is equally essential. The latter is usually more challenging, although the web, search engines, and various technology-enabled presentational techniques have made it both easier and more powerful. The results can be vivid, as the pioneering work of Hans Rosling and collaborators with animated data illustrates – see Rosling and Rosling (2006), Rosling (2010), Rosling et al. (2018) and the Gapminder website [1H]. We shall discuss getting meaning out of data in more detail in Section 1.5.

Note that the data-driven approach tells you nothing about the mechanisms that underlie the phenomena. For that you need to turn to modelling – mathematical or purely descriptive. This dialectic between these two complementary approaches, data-driven and theory-driven, is at the heart of understanding phenomena – and of mathematical literacy.

Critical thinking and critical enquiry

Critical thinking as outlined above is essential for all mathematical literacy. However, it is especially key to approaches to theorising mathematical literacy from the perspective of critical enquiry (see e.g. Jaworski, 2006). In this work, contexts will generally be personal or societal, often with substantial political implications that concern us as citizens. In this 'critical thinking' approach, the role of mathematical knowledge is frequently mainly as a means of presenting data, and providing support for asking critical questions – such as whether the right variables have been selected, whether the methodology for data collection and interpretation is sound, and what the real-world implications are.

The theoretical model of numeracy developed by Goos et al. (2019) emphasises a critical orientation – by which they mean "the use of mathematical information to: make decisions and judgments, add support to arguments or challenge an argument or position". Like our framework, their model puts contexts at the centre, and recognises the importance of mathematical knowledge and a productive disposition to thinking mathematically, and they add the importance of mathematical tools.

Command of language, more important in mathematics than is sometimes recognised, is central to mathematical literacy, especially to achieving a critical orientation – for the analytic discussion of data, the issues it suggests, and for the construction of powerful, logical arguments. Mathematics here may be very much in a supportive role – but a potentially powerful one.

One strand of the diverse critical thinking community in mathematics education (see e.g. Andersson & Barwell, 2021) focuses on global concerns, some of which are the focus of later chapters here, with a socially proactive approach that "is interdisciplinary; is politically active and engaged; is democratic; involves critique; and is reflexive and self-aware" (p. 7).

In the next two sections we examine the processes of mathematical modelling and of getting meaning out of data in a bit more depth.

1.4 Modelling real situations with mathematics

How do we set about using our mathematics to better understand a practical situation? This section develops the 'theory-driven' *know how* aspect of mathematical literacy, particularly taking the context seriously, possessing a range of appropriate mathematical and techno-mathematical skills, and being able to deploy them in understanding the situation. The long-established term 'mathematical modelling' became part of everyday language only during the 2020–2022 COVID-19 pandemic, when the predictions of the various modelling groups were hot news. 'Modelling' has generally been thought to need advanced mathematics; yet like Monsieur Jourdain, Moliere's *Bourgeois Gentilhomme*, who was surprised to learn that he had been speaking *prose* all his life, most people have been modelling with mathematics at a basic level for many years – predicting the outcome of sharing sweets as young children, working with money, planning an event.

The understanding of the modelling process and its phases, as depicted in Figure 1.3.1, is not essential for using mathematics to tackle real-world problems but is valuable for teachers in promoting awareness of the process and for self-monitoring when tackling problems.

Various rather different activities are involved in modelling, moving from the practical situation into a mathematical representation of some aspects of it and back to the situation. Let's talk it through using the task *Sauce* in Figure 1.4.1 (assuming for the moment that it is a novel task rather than a learned procedure for the solver). Even in a solution to this simple and rather stylised task, the phases of modelling, as set out in Figure 1.4.2, are evident.

You are making your own batch of dressing for salads.

Here is a recipe for 100 millilitres (mL) of dressing taken from the web.

Recipe for salad dressing	
Salad oil:	60 mL
Vinegar:	30 ml
Soy sauce:	10 mL

Q1: How many millilitres (mL) of salad oil do you need to make 250 mL of this dressing?

FIGURE 1.4.1 Sauce.

Source: Adapted from PISA 2012 released mathematics item [1C].

The modelling process	The outcome
It is important to get to understand both the **problem in its context and the task** qualitatively. What are the important quantities, the *variables*, and how are they *related*. What is the question – the task in hand? Read the *Sauce* task carefully, looking for the significant quantities and the results asked for. *(In designing assessment tasks like this PISA mathematics task, every effort is made to make this phase easy.)*	A **conceptual model** of the situation and the task. This sauce is a mixture of three liquids and in this case the volumes simply add (not always true). The task is to work out the scaled-up volumes of the ingredients, given the new total.
Next **formulate** a mathematical representation of the situation and the task. The variables are the volumes of each ingredient and the total in the recipe. The task is to represent the relationships when the total volume is changed from 100 mL to 250 mL.	A **mathematical model** of the situation and the task. To change the total volume, all ingredient volumes are multiplied by the scale factor, the ratio of the total volumes.
Solve the problem mathematically by manipulating the model to calculate the quantities the task specifies. Calculate the scale factor and carry out the multiplication for each ingredient. Other calculation methods are possible.	A **mathematical result** $250/100 = 2.5$ $60 \times 2.5 = 150$ $30 \times 2.5 = 75$ $10 \times 2.5 = 25$
Interpret the mathematical result in the context of the practical situation. Link the mathematical answers to each ingredient.	The **result in context** Oil 150 mL, vinegar 75 mL, soy 25 mL
Evaluate the results in the context of the situation. Do they make sense? *If so, use or communicate* the result. *If not*, check the mathematics; if that seems correct, go back to the beginning and *improve the model* For *Sauce* the model seems to work well *but* we have not looked at how well different quantities emulsify – but this probably would not change between 100 mL and 250 mL.	These are **either** useful results *or* some insights to guide improving the model. You can use these quantities, *but if the desired volume was much greater*, you may need to mix them in small quantities to ensure they emulsify properly.

FIGURE 1.4.2 The phases of modelling illustrated by *Sauce*.

We shall have much more to say about modelling in action in later chapters. A fuller discussion and analysis, including the development of these ideas in various forms, can be found in Stacey and Turner (2015) and in another book in this series: *The Learning and Teaching of Mathematical Modelling* (Blum & Niss, 2020). Teaching materials focused on the modelling process can be found in the reSolve Special Topic unit on Mathematical Modelling [1D] (examples of which can be found in Chapters 4 and 10 in this book).

What is the value of this analysis of the modelling process, this 'model of modelling', in developing mathematical literacy? In any process that requires thought, it is usually helpful to be aware of what you are doing; particularly when you get stuck and wondering what to do, such metacognitive reflection helps to 'break the log jam'.

- In *formulating* the mathematical model, usually the most challenging phase, you may need to ask yourself questions like: Have I chosen the right variables for what I'm trying to understand? What have I missed out that is significant? Do I really need to consider this? How are these quantities related? How can I represent these relationships? Should I look at a particular case in more detail? Would a spreadsheet help me to explore possible ways forward in the light of the solutions that emerge? Do I need further information? Some contexts are so important that students are *taught* the appropriate models (e.g. to increase quantities in proportion, relationships between distance, speed, and time, or for money conversion). In others the relationships need to be constructed from first principles when a new problem situation is tackled.
- The *solve* phase is well covered by school mathematics – though in non-routine problem-solving, students can, in practice, only use concepts and skills that they have thoroughly understood and connected to other parts of mathematics and to a variety of contexts. Of course, it is also possible that a solution to the mathematical problem is not available within the solver's level of mathematics; in which case, there is no alternative but to formulate in a different, possibly simpler, way. Or perhaps numerical methods using a graphing calculator or a spreadsheet might enable a solution. Many of the complex models used by experts are only solvable in these ways.
- After the thinking done for the formulation phase, the *interpret* phase is usually relatively straightforward, reflecting what was considered in formulating the model. But the results may be surprising, with potentially interesting life-related implications, or just wrong.
- For the *evaluate* phase, we are back in the real world, thinking about the practical situation. Does the result make sense and does it provide a useful answer to the question? If it looks dubious, check the mathematics again. If it still doesn't look right, do I need to formulate the problem in another way?

We shall face questions like these in the examples throughout this book and beyond. To summarise, the modelling diagram of Figure 1.3.1 can be helpful both as a guide and as a checklist.

Finally, a word of caution. It is important to avoid unreasonable expectations about the answers that a model can give – in particular, in predicting the future. For example, modellers' analyses of the COVID-19 pandemic received a lot of publicity, with people expecting answers to 'small' details. Should a lockdown limit people to 5 km or 10 km? Should people have to wear a mask when outside? In fact, the model predictions were not 'accurate'; they were used by sensible people as the best guides available, month by month, in a situation with many unknowns. The differences between different modelling groups gave some indication of the uncertainties. A model is only as good as the assumptions and the data that go into it which, for real-life systems, are rarely complete and accurate. Hence, again, "All models are wrong but some are useful" – in itself inexact but useful guidance. (See, for example, Pablo Rodríguez-Sánchez's blog entry *Can mathematical models predict the future?* [1E].)

Standard models versus active modelling

In taking the context seriously, there is a fundamental distinction that is important to bring out – between *standard models* or *applications* and *active modelling*. When a new topic is introduced in school mathematics, it is often accompanied by examples of that topic being used in various practical situations – for example, adapting a recipe as an application of proportional reasoning in *Sauce,* above. These *standard applications* are models of everyday situations that the student is taught, primarily as examples of the mathematical topic in use in common situations of some importance. They serve:

- to reinforce understanding of the concept and develop skills through a concrete embodiment,
- to equip students with models that describe some very common real-world situations, along with
- the hope the student will recognise other situations where similar mathematical relationships are involved.

However, situations frequently arise in everyday life that have not been covered in the classroom (or have been forgotten) but which people can better understand by using their mathematics. Such situations require *active modelling*. Figure 1.4.3 illustrates these very different mathematical activities.

Learning standard models and active modelling are complementary activities – both are important in developing mathematical literacy. In teaching standard applications, the emphasis is on students' mastery of the new

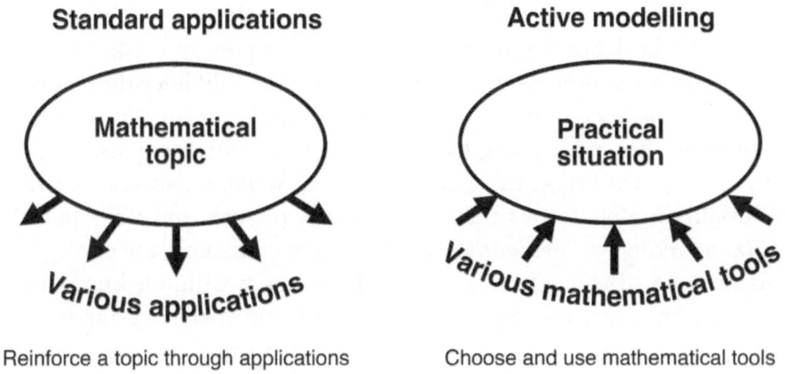

FIGURE 1.4.3 Standard applications versus active modelling.

mathematical topic and canonical examples, which usually depict simplified and stylised real-world situations. Tackling new situations, which demand understanding the practical situation in some depth, requires active modelling, using whatever mathematical tools seem likely to be useful in that process. Learning standard applications is important in building a toolkit of models that can be adapted to new situations as they arise. But recognising possibilities in formulating a model is helped by learned connections to a wider variety of situations than are usually brought in when the mathematical topic is the focus. Building these connections is at the heart of developing mathematical literacy. We can see this in Figure 1.4.4 from *Be a Paper Engineer* – part of the *Numeracy through Problem Solving* materials discussed in Section 2.4.

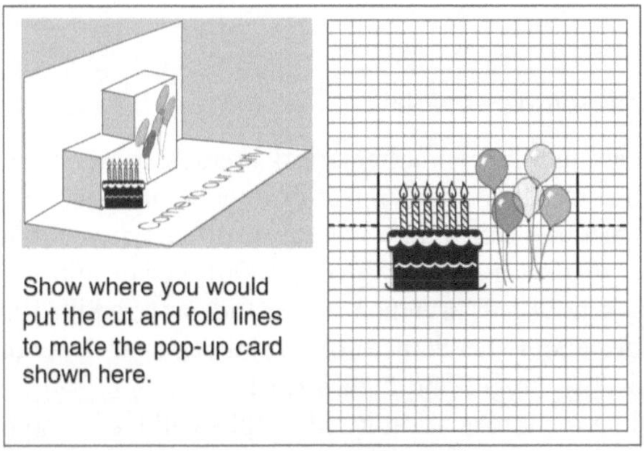

FIGURE 1.4.4 Design and make a party invitation.

Source: Be a Paper Engineer – Shell Centre for Mathematical Education [1F].

To design the party invitation, *formulation* involves recognising that the lengths of the horizontal and vertical sides of the pop-up are key *variables*, and seeing their *relationship* to the pictures and the fold lines they show. For the relationships, recognition of the length equalities is key – essentially the parallelogram theorem on lengths. (Only a few students recognise it as such; for others, the card brings the theorem to life.) *Solving* involves transferring these insights in detail onto the net, using the pictures for scale and alignment. *Interpreting* and *evaluation* are best done by making the card.

Finally, a word of caution – again. The purpose of modelling is to give some useful insights. But no model is perfect and many are far from it – assumptions may be inadequate and/or significant variables may be ignored. For example, money models often seem perfect because the whole situation is a mathematical construct – the rules for adding interest to a savings account are defined mathematically in a contract; but this model takes no account of how the interest rate may be changed in the future. Moreover, it takes no account of inflation which historically has meant that the real interest rate is negative – you are actually paying people to borrow your money; estimated corrections for that effect, called a 'net present value', are of course inexact. In the end, the evaluation phase requires judgement based in real-world understanding and further critical reflection on what to do next. We look at this and other aspects of financial literacy in Chapter 9.

This book is focused on modelling to better understand the real world, with known mathematics. The Realistic Mathematics Education (RME) project, developed at the Freudenthal Institute (see e.g. Van Den Heuvel-Panhuizen, 2003), takes modelling as an approach to teaching and learning: starting with carefully chosen real-world or fantasy-world situations from which mathematical concepts and representations can be abstracted and developed. The two approaches are complementary and mutually supportive.

1.5 Getting meaning out of data

The 'data-driven' aspect of mathematical literacy comes in when you are presented with some data on a subject of interest or concern: starting from this point is sometimes called *data literacy* or often, where random variation is important, *statistical literacy* (Watson, 2006). For example, a report on a bus crash on a school trip in which some students are killed might make a teenager wonder about the chances of a fatal accident happening. Looking for data on this, they might find Figure 1.5.1.

Understanding the information in the graphs will lead to some inferences. The chance of a young person dying in the coming year is somewhere between 1 in 10 000 and 1 in 1 000. Even though the probabilities are very low, from age 15 they are a lot higher for boys than for girls. This leads to

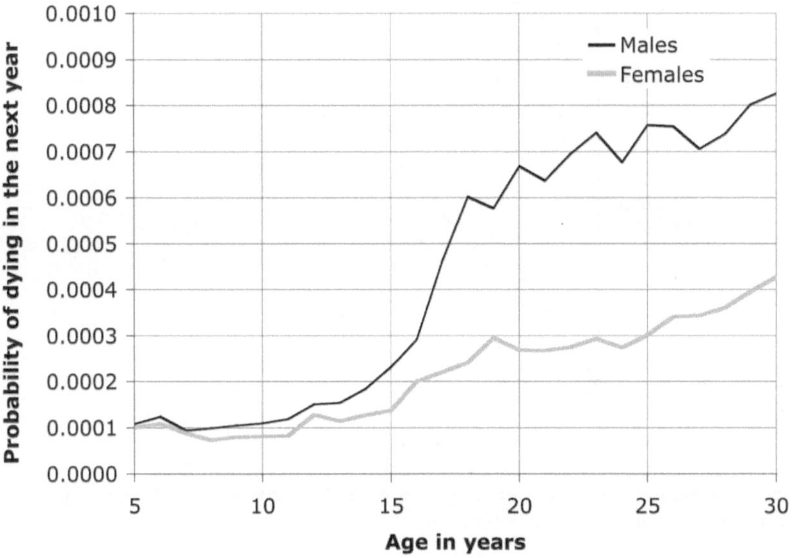

FIGURE 1.5.1 The average risk of dying in the next year.

Source: From Bowland Maths – *How risky is life?* [1J].

the immediate questions "Why?" then "What are the main causes of death, and of the gender differences?" This leads onto the search for more data and, sometimes, for models that explain it. We return to look at risk and the *How risky is life?* materials in Chapter 4.

This is a typical example of the processes of *data literacy*. Starting from a data-based mathematical description, in this case two graphs and their axes, it resembles the *interpretation* and *evaluation* phases of the modelling process: understanding the meaning of information presented in various forms – generally tables of numbers or graphs of various kinds – and seeing how appropriate the data is to the situation of concern. Sometimes this leads to trying to relate the data to underlying causes through seeking further data or active modelling.

In a school context, this distinction is illustrated by the tasks *Design and Make a Party Invitation* above, which involves the full modelling process, and *Hurdles Race* (Figure 1.5.2) which just requires the interpretation and interrogation of data presented as a graph. *Hurdles Race* (from Swan et al. 1985, [1G]) is challenging in the following ways. The insights have to be assembled into a coherent real-time narrative in the style of a radio commentary. Giving a commentary from a graph like this is not going to arise in the real world, but it is a valuable exercise in getting meaning out of graphs, a key part of data literacy, as well as reinforcing the concepts involved. *Hurdles Race* thus illustrates the mutual support that data interpretation and conceptual

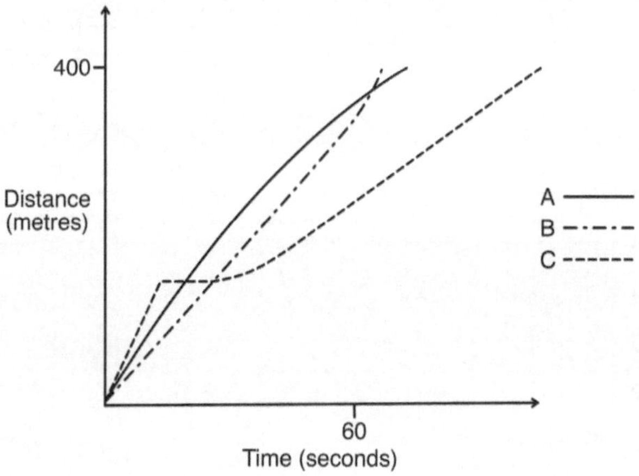

The rough sketch shown above describes what happens when 3 athletes A, B and C enter a 400 metres hurdles race.

Imagine that you are the race commentator. Describe what is happening as carefully as you can. You do not need to measure anything accurately.

FIGURE 1.5.2 Hurdles race.

Source: From *The Language of Functions and Graphs.* Swan et al. 1985 [1G].

understanding can provide – a win-win for both mathematical literacy and mathematical understanding (Figure 1.5.2).

Figure 1.5.3 gives several examples of data arising from real situations of wide interest which are discussed in later chapters of this book. The first challenge is to interpret the graphs, identifying the variables and scales and features of the data that are interesting. The next is to discuss the implications – individual, societal, and political – of what is revealed. The third is to begin to list further questions that arise and investigate where data may be found to illuminate them. Finally, it is sometimes worth exploring whether a mathematical model might be constructed that would offer an analytic explanation and enable predictions.

The last 50 years have seen the development of very powerful ways, initially called 'exploratory data analysis' (Tukey, 1977), in which technology can help us transform complex data into much more effective forms. Hans Rosling (2010), in particular, developed ways of showing on screen many kinds of historic population and human data and going beyond the traditional two variables by using colour and animation. That this work has impacted mainstream media is a tribute to its power in getting meaning out of data. In later chapters, we shall be concerned with data arising in a diverse range of specific situations.

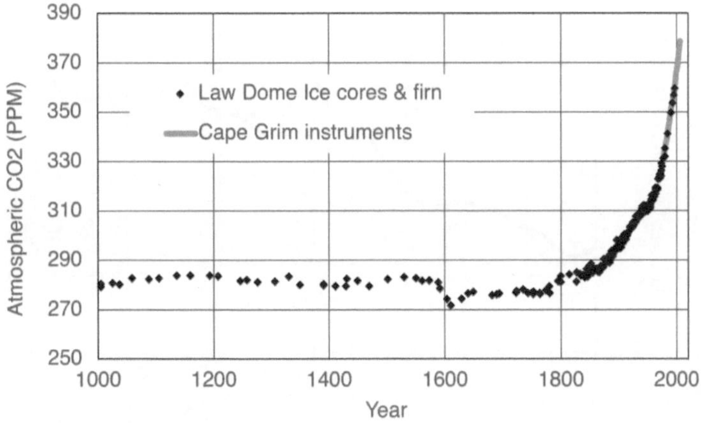

(a) Atmospheric CO2 levels since AD 1000. Data from NOAA – see Chapter 5.

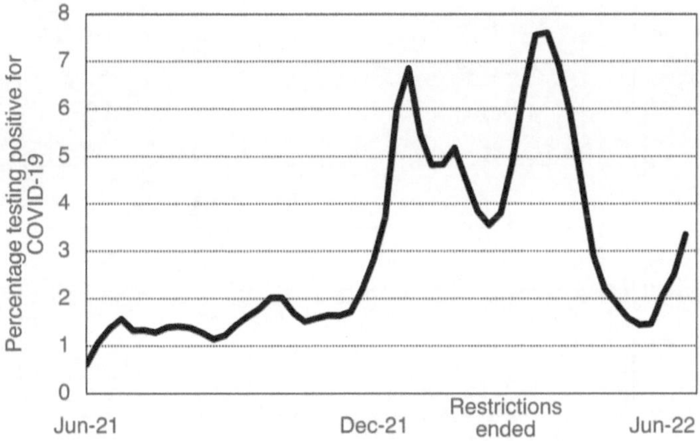

(b) COVID-19 infection levels in England over time. Data from UK ONS – see Chapter 4.4.

FIGURE 1.5.3 Data for classroom analysis and discussion. *(Continued)*

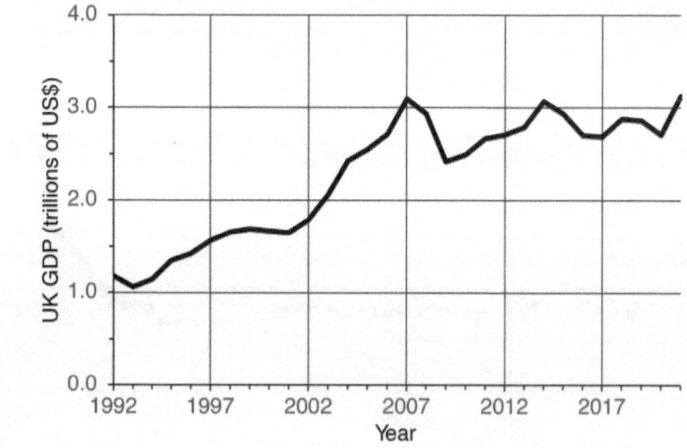

(c) UK gross domestic product 1992–2021. Data from World Bank – see Chapter 7.3.

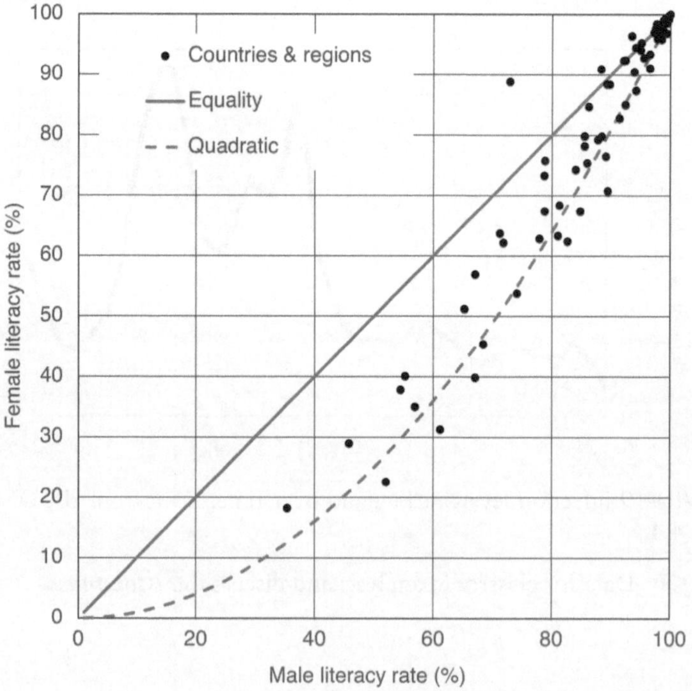

(d) Male versus female literacy (%) across countries. Data from UNESCO – see Chapter 10.5.

FIGURE 1.5.3 *(Continued)*

1.6 The power of technology

In the world beyond school, most mathematical activity involves computer-based platforms of various kinds – from calculators to cash registers in shops, through mobile phones to supercomputers. That is how nearly all calculation, information processing and data presentation is done nowadays. This reality has been slow to penetrate school mathematics classrooms for multiple reasons. For a long time, concerns about equality of provision between different schools and different students led curriculum and examination authorities to exclude technology. As ever, this influenced teachers and textbook writers. There is also a view that progress in learning mathematics is covering more advanced concepts but always exemplified only through simple tasks where technology is less valuable. Further, curriculum time is always tight, so teaching students to use technology can be seen as an extra burden. The situation has improved steadily in this century, but technology-rich curricula remain the exception.

Most people still leave school with an acquaintance with a lot of mathematics that they cannot, in practice, apply to more than the simplest situations. Indeed, most adults use little if any of the mathematics that they first meet after about age 14. The rapid, accurate execution of well-defined usually numerical procedures – which gained many people useful employment in the distant past – is now the domain of technology. Yet there still seems to be a lack of clarity over what needs by-hand mastery and what can be safely left to technology. Mathematical literacy offers at least a partial justification for the large amount of curriculum time that is still spent on mathematics but, for this, it must be allied with technology. How?

The core of the approach is an alternative view of progress in mathematics – as the ability to tackle *increasingly complex tasks* using increasingly sophisticated concepts and tools. (This is analogous to a widely accepted definition of progress in first language studies – with *tasks* replacing *texts*.) *Task* complexity – as opposed to the sophistication of the concepts – is the new element. The real-world problems that mathematical literacy embraces tend in that direction (as do those in really doing abstract mathematics – but that's beyond our brief here).

A simple example is solving quadratic equations. There is a famous formula for the two solutions with its important and elegant proof, but too many students forget it quickly and some never understand it. Teaching about the quadratic formula introduces some widely applicable algebraic techniques, but the formula itself only applies to quadratic equations. However, with easy calculation using technology, guess–check–improve techniques are viable ways of solving equations – not elegant and not exact but easy to carry out to as many decimal places as required, easy to remember, and applicable to many types of equations, not just quadratics. Numerical methods for solving equations with technology, refined to make them robust and fast, are routinely used in industrial, engineering, and scientific work.

Mathematically literate citizens should at least know about this. Mathematics students should know how to use basic guess–check–improve methods effectively and to use simple tools such as spreadsheets or calculators to make them viable. Later chapters contain many examples of technology-supported methods.

Simulation

Many systems are too complex to find mathematical relationships that describe them adequately, but they can be explored using computer-based 'experiments'. In a simulation, analytic assumptions, including statistical variability, are fed in at a 'micro-level' and subsequent macro-level behaviour is calculated. This is repeated many times with minor changes to the initial conditions and/or the parameters so as to understand the behaviour of the system. Simulation can also be used in systems where, though the underlying laws are completely understood and deterministic, the system behaviour is very complicated or sensitive to minor changes in the initial conditions or parameters. Simulation is now so important that it stands alongside theory and experiment as one of the 'pillars of science'.

Weather forecasting is a well-known example where large computer programs are used to simulate the atmosphere. Millions of individual observations of the atmosphere are obtained from satellites and a global network of thousands of weather stations. The data is fed into the model, and the future weather is simulated, running faster than it happens in reality. In these systems, small variations in the data that are well within the uncertainties of the observations can produce big changes in the predictions. This is fancifully called 'The Butterfly Effect': a butterfly flaps its wings in one place and produces a tornado far away. To make forecasts more reliable, weather forecasters run their calculations (simulations) many times with minor changes in the input data. They then look for common features in the predicted weather patterns – these features are used for the weather forecast. But it is expected that occasionally one of the unusual patterns will actually happen.

1.7 The nomenclature problem

The book has a broad and ambitious interpretation of the term *mathematical literacy*. However, other narrower uses of the term are common and it is important to be alert for the distinctions. Just as *literacy* is used in English both for a broad knowledge of literary culture (being 'well read', etc.) and for the narrow ability to read and write at a basic level (as opposed to *illiterate*), both *mathematical literacy* and, particularly, *numeracy* are often narrowed down to mean just competence in the basic skills of arithmetic. This ambiguity has undermined the development of mathematical literacy in many education systems. Even when mathematical literacy is an official goal, it is often

- A pair of jeans is on sale at "20% off" for \$40. What was the original price?
- In a grocery store, 4 kg of tomatoes cost \$5. How much will 7 kg cost?
- t seconds after an object falls from a height of H metres, its height h in metres is given by the formula $h = H - 5t^2$

  If H is 20 metres, when will the object hit the ground?

FIGURE 1.7.1 The narrow meaning for numeracy or mathematical literacy.

implemented narrowly as simply emphasising a subset of traditional school mathematics, mainly arithmetic. This familiar material is easier to teach and to assess than mathematical literacy as described here. But without the "capacity to formulate, employ, and interpret mathematics in a variety of contexts" – from the PISA definition – such skills are not enough to enable people to use their mathematics to help with challenges of everyday life, or to 'know about' enough mathematics to satisfy their curiosity about other topics of interest.

Figure 1.7.1 exemplifies the narrow interpretation. All of the tasks there might be described as "using mathematics to tackle a problem in everyday life", but they draw on only the most minimal understanding of the contexts. The only discussion of context likely to happen is asking students who are having difficulty with the third task: "What does 'hit the ground' mean about the height?"

In contrast, the *Airplane Turnround* task in Figure 1.3.2 really does demand thinking about the real-world situation. When students recognise that the task requires considering which of the jobs can be done in parallel, they can come up with a shorter time. There should be discussion about the various assumptions being made – is it safe to refuel the plane when passengers are on board? – and recognition that they would need more information to make a fully informed decision. This is an example of "the sophisticated use of elementary mathematics" (Steen, 2002; Steen & Forman, 2000) that distinguishes mathematical literacy from skill exclusively within mathematics. The computational skill in this task is not demanding; the difficulty arises instead from the need to work with both the situation and the mathematics – critical thinking about the mathematical structure of turnround jobs done in parallel or series, as well as the modest 'technical skill' of calculation.

That ambiguities of meaning arise is not surprising. (Mathematical literacy is also occasionally used in ways entirely unconnected with everyday life – as a broad knowledge of pure mathematics in all its specialties, for example.) The descriptive use of language is often ambiguous. Particularly when something unfamiliar is described, people naturally interpret the words within their prior experience. But there is a strategy that sharply narrows such ambiguities: *making meaning clearer through examples*. We shall do so throughout this book.

Using the term mathematical literacy in the broad sense of the PISA definition (and including, as noted above, statistical and other cognate literacies), the obvious next question: "How broad?" The tasks in the next chapter, and in those that follow begin to provide the answer. In presenting them, we will also point to the essential roles played by the concepts, reasoning structures, and the technical skills of abstract mathematics (by technical skills, we mean skill in calculating, doing algebra, etc.). These are immediately clear to teachers in the tasks in Figure 1.7.1; they are equally essential in tackling life-related tasks.

Another source of confusion when describing mathematical literacy is that the versions that are valuable and feasible for different people differ immensely in complexity. An adult working in a scientific or financial field will need a much higher level of mathematical literacy than a young person just starting out in the music industry might. We can say that both need mathematical literacy in their work and personal lives, but the level of mathematics used and the nature of the real-world situations they encounter are very different. In this book, we focus on examples of mathematical literacy that draw on typical school mathematics and involve life-related situations that are likely to be of interest to many secondary school students in their current lives and when looking forward. We also include some examples that arise in more specialised work situations that are of wider significance – in finance, for example.

For many problem situations, mathematical literacy involves this more sophisticated use of simple-but-robust mathematics, particularly arithmetic, proportional reasoning, and graphs. Along with resources that technology offers, these provide the tools a mathematically literate person needs for both simple modelling of practical situations with mathematics and 'getting meaning out of data'.

1.8 Summary and forward look

In this chapter, we have set out to clarify the meaning of mathematical literacy through descriptions, definitions, and the context-focused mathematical framework built from five broad components of mathematical literacy. We have analysed the processes involved in modelling real situations with mathematics. We have begun to explore the processes of getting meaning out of data. We have illustrated the power of technology and numerical methods in both these areas. It is clear that mathematical literacy involves a deliberative approach to thinking about a practical problem – what Daniel Kahneman (2011) called "slow thinking" – critically questioning one's instinctive, sometimes emotional initial reactions. The challenges and opportunities for teaching that mathematical literacy presents are the focus of Chapter 3 – but first we enrich the analysis of this chapter through a diverse range of examples of classroom teaching for mathematical literacy.

Acknowledgements

Figure 1.4.1 is an adaptation of an original work by the OECD (PISA 2012 released mathematics item PM924Q02). The opinions expressed and arguments employed are the sole responsibility of the author or authors of the adaptations and should not be reported as representing the official views of the OECD or of its member countries. PISA materials are licensed under the Creative Commons Attribution-NonCommercial-ShareAlike 3.0 IGO (CC BY-NC-SA 3.0 IGO) licence. Figures 1.4.4 and 1.5.2 are materials from the Shell Centre for Mathematical Education and appear courtesy of the Bell Burkhardt Daro Shell Centre Trust. Figures 1.3.2 and 1.5.1 appear courtesy of the Bowland Maths maintainers.

References

Andersson, A., & Barwell, R. (2021). *Applying critical mathematics education.* Brill.

Box, G. E. P. (1979). Robustness in the strategy of scientific model building. In R. L. Launer & G. N. Wilkinson (Eds.), *Robustness in statistics* (pp. 201–236). Academic Press. https://doi.org/10.1016/B978-0-12-438150-6.50018-2, ISBN 9781483263366.

Burkhardt, H. (2008). Quantitative literacy for all. In B. L. Madison & L. A. Steen (Eds.), *Calculation vs context: Quantitative literacy and its implications for teacher education* (pp 137–162). Mathematical Association of America. Retrieved May 19, 2023 from https://www.maa.org/sites/default/files/pdf/QL/cvc/CalcVsContext.pdf

Burkhardt, H. (2018). Ways to teach modelling: A 50 year study. *ZDM*, *50*(1), 61–75. https://doi.org/10.1007/s11858-017-0899-8.

Crowther Report 15–18. (1959). *A report of the Central Advisory Council for Education.* HMSO.

Geiger, V., Goos, M., & Forgasz, H. (2015). A rich interpretation of numeracy for the 21st century: A survey of the state of the field. *ZDM Mathematics Education, 47*, 531–548. https://doi.org/10.1007/s11858-015-0708-1

Goos, M., Geiger, V., Dole, S., Forgasz, H., & Bennison, A. (2019). *Numeracy across the curriculum: Research-based strategies for enhancing teaching and learning* (1st ed.). Allen & Unwin.

Jaworski, B. (2006). Theory and practice in mathematics teaching development: Critical inquiry as a mode of learning in teaching. *Journal of Mathematics Teacher Education*, *9*(2), 187–211.

Kahneman, D. (2011). *Thinking, fast and slow.* Farrar, Straus and Giroux.

Madison, B. L. and Steen, L. A. (2008) *Calculation vs context: Quantitative LITERACY and its implications for teacher education* (pp. 137–162). Mathematical Association of America. Downloaded as https://www.maa.org/sites/default/files/pdf/QL/cvc/CalcVsContext.pdf

National Governors Association (2010) Common Core State Standards in Mathematics. Washington, DC

Niss, M., & Blum, W. (2020). *The learning and teaching of mathematical modelling.* Routledge.

Organisation for Economic Co-operation and Development (OECD). (2013). *PISA 2012 assessment and analytical framework: Mathematics, reading, science, problem solving and financial literacy*. OECD Publishing. https://doi.org/10.1787/9789264190511-en

Rosling, H. (2010). *Global population growth, box by box* [Video]. TED Conferences. https://www.ted.com/talks/hans_rosling_global_population_growth_box_by_box

Rosling, H., & Rosling, O. (2006). How not to be ignorant about the world. Gapminder. Sweden. Retrieved May 19, 2023 from https://policycommons.net/artifacts/2475074/how-not-to-be-ignorant-about-the-world-gapminder/3497131/. CID: 20.500.12592/s5h5hz.

Rosling, H., Rosling, O., & Rosling Rönnlund, A. (2018). *Factfulness: Ten reasons we're wrong about the world: And why things are better than you think*. Flammarion.

Stacey, K., & Turner, R. (2015). *Assessing mathematical literacy: The PISA experience*. Springer.

Steen, L. A. (Ed.). (2002). *Mathematics and democracy: The case for quantitative literacy*. National Council on Education and the Disciplines.

Steen, L. A. (1999). Numeracy: The new literacy for a data-drenched society. *Educational Leadership*, *57*(2), 8—13. Retrieved May 19, 2023, from https://www.ascd.org/el/articles/numeracy-the-new-literacy-for-a-data-drenched-society

Steen, L. A., & Forman, S. L. (2000). Making authentic mathematics work for all students. In A. Bessot & J. Ridgway (Eds.), *Education for mathematics in the workplace* (pp. 115–126). Kluwer.

Steen, L. A., Turner, R., & Burkhardt, H. (2007). Developing mathematical literacy. In W. Blum, P. L. Galbraith, H.-W. Henn, & M. Niss (Eds.), *Modelling and applications in mathematics education*. the 14th ICMI study. Springer.

Swan, M. with Pitts, J., Fraser, R., and Burkhardt, H, and the Shell Centre team (1985). *The language of functions and graphs*. Shell Centre for Mathematical Education. https://www.mathshell.com/materials.php?item=lfg&series=tss

Tukey, J. W. (1977). *Exploratory Data Analysis*. Adddison-Wesley.

Van Den Heuvel-Panhuizen, M. (2003) The didactical use of models in realistic mathematics education: An example from a longitudinal trajectory on percentage. *Educational Studies in Mathematics*, *54*, 9–35.

Watson, J. M. (2006). *Statistical literacy at school: Growth and goals*. Lawrence Erlbaum.

Links to useful material

To visit any of these links, scan this QR code or visit ltml.mathlit.org – append the link code to go directly to the entry – for example, ltml.mathlit.org/1A

The original source links are given below for attribution purposes:

[1A] *PISA 2022 Mathematics Framework* – OECD
https://pisa2022-maths.oecd.org/ca/index.html

[1B] *Programme for the International Assessment of Adult Competencies* (PIAAC) – OECD
https://www.oecd.org/skills/piaac/piaacdesign/
[1C] *PISA Released Mathematics Items 2012* – OECD
https://www.oecd.org/pisa/
https://www.oecd.org/pisa/pisaproducts/pisa2012-2006-rel-items-maths-ENG.pdf
[1D] *Special Topic: Mathematical Modelling* – reSolve
https://www.resolve.edu.au/mathematical-modelling
[1E] *Can Mathematical Models Predict the Future?* – Pablo Rodríguez-Sánchez
https://blog.esciencecenter.nl/can-mathematical-models-predict-the-future-c362a0fbced2
[1F] *Be a Paper Engineer* – Shell Centre for Mathematical Education
https://www.mathshell.com/materials.php?series=numeracy&item=paperengineer
[1G] *The Language of Functions and Graphs* – Shell Centre for Mathematical Education
https://www.mathshell.com/materials.php?item=lfg&series=tss
[1H] Gapminder
https://www.gapminder.org/
[1J] *How Risky Is Life?* – Bowland Maths
https://www.bowlandmaths.org /projects/how_risky_is_life.html

2

THE POWER OF TASKS

We now move on from the general description and discussion of mathematical literacy to show something of the *variety of tasks* that learning mathematical literacy involves, presenting and analysing task examples of different kinds and lengths. Why the focus on tasks? We have seen in Chapter 1 how illustrating descriptions with task examples clarifies the kind of thing one is talking about. As we noted in Section 1.6, there is another, deeper reason to give tasks a central role:

> Progress in learning mathematics, and mathematical literacy, can be thought of as the ability to tackle increasingly complex tasks using increasingly sophisticated concepts and skills.

We aim to use the power of examples to build a down-to-earth 'universe of discourse' about mathematical literacy, leading into a discussion of teaching, pedagogy, and practice in Chapter 3.

All the tasks that follow work well in classrooms. Many have been chosen because they come from teaching materials that have been imaginatively designed, carefully developed in classroom trials and are available online. Such 'well-engineered' materials will be referenced throughout the book, often with direct links, as resources for both teacher education and school classrooms. The emphasis here is on 'know how', on students learning to solve life-related problems with mathematics, illustrating ways in which the goals of developing mathematical literacy have been forwarded by teachers.

DOI: 10.4324/9781003303503-3

It is useful to identify some broad 'task types' that characterise mathematical literacy and are shown in the examples that follow. Tasks may ask students to do one of the following:

- *Model and explain:* invent, interpret, and explain models to understand real-world phenomena and make predictions; modelling, as we have described in Chapter 1, is an important component of mathematical literacy.
- *Plan and organise an event or a programme:* find an optimum solution for the goals of the event, subject to constraints of time, space, money, etc.
- *Design and make an artefact or procedure, and test it:* a common real-world activity; the importance of thinking about *imaginative design* is often overlooked.
- *Explore and discover:* finding relationships, often from looking at data in some depth, leads on to hypotheses as to underlying causes, which may later be expressed as models with predictions.
- *Interpret, and translate between, representations:* this is done with words, numbers, graphs, or algebra to extract meaning from data, or from model calculations.
- *Evaluate and improve:* review and improve an argument, a plan, or an artefact. This kind of activity is less challenging than *creating* an argument, plan, or artefact and can smooth the entry to a new context.

The aim of this list is not to give water-tight definitions that definitively classify tasks (you might disagree with some of the classifications below), but to help ensure that important kinds of mathematical literacy activity are not overlooked in planning a mathematical literacy element in a curriculum. All five components of the context-focused mathematics framework are important in each of these areas.

The majority of tasks in this chapter require at least some parts of the theory-driven mathematical modelling cycle. Later chapters will present many data-driven problems. The mathematical techniques needed for the tasks used here are standard (although not necessarily easy for students), but each task also requires some strategic thinking especially related to the formulating and evaluating phases of modelling discussed in Sections 1.3 and 1.4. This involves questions such as:

- Formulate
  - How might I tackle this problem?
  - How might I organise and represent the situation?
  - Which bits of the mathematics I know could be useful?
  - What assumptions do I need to make? What are the important variables?
  - What are the mathematical relationships, if any, between the variables?

- Evaluate
  - Do my answers seem reasonable, given what I know of the real-world situation?
  - How can I justify my conclusion, explaining my reasoning?

This last point, reflected in the phrase "to describe, explain and predict phenomena" from the PISA definition of mathematical literacy given in Chapter 1, reflects the importance of students' explanations and justifications – in mathematical literacy and, equally, in really 'doing mathematics'.

2.1 The range of tasks for mathematical literacy

We begin with the two simple tasks from PISA surveys in Figure 2.1.1. Both could be used during the teaching of proportional reasoning.

Space flight is a *model and explain task*, built on recognising that proportional reasoning applies – a scaling problem of a kind that arises regularly in everyday life and work. *Sauce* in Figure 1.4.1 and the tomatoes task in Figure 1.7.1 are other examples. Mathematically, there are several lines of correct reasoning in '3-number proportion' problems like this. Conceptually,

Space Flight

Space station Mir remained in orbit for 15 years and circled Earth some 86 500 times during its time in space.

The longest stay of one cosmonaut in the Mir was around 680 days.

Approximately how many times did this cosmonaut fly around Earth?

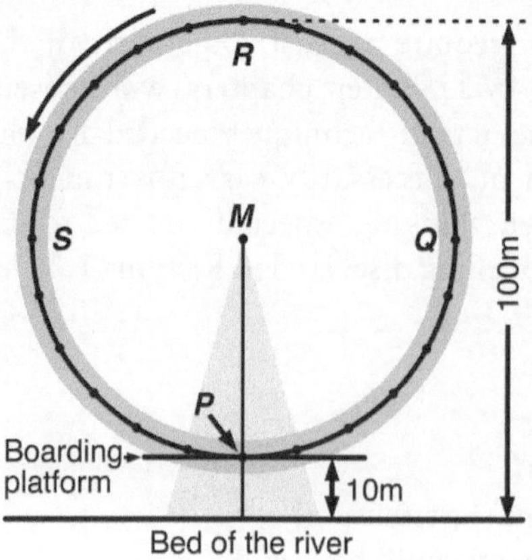

Ferris Wheel

A giant Ferris wheel is on the bank of a river

See the diagram to the left

The Ferris wheel rotates at a constant speed. The wheel makes one full rotation in exactly 40 minutes

Q2: John starts his ride on the Ferris wheel at the boarding point, *P*. Where will John be after half an hour?

FIGURE 2.1.1 Two short mathematical literacy questions.

Source: PISA Released Mathematics Items [2A].

perhaps the most straightforward is to recognise a scaling situation, calculating the *scale factor* from the two given *quantities that have the same dimensions*, in this case *time*. Converting both times to days, we get a scaling factor to multiply by the total number of orbits, giving

$$\frac{680}{15 \times 365.25} \times 86\ 500 \approx 10\ 736$$

Ans: The cosmonaut flew around Earth about 11 000 times. The PISA item indicated the accuracy required by the multiple-choice options supplied: {110, 1100, 11 000, 110 000}. A mathematically literate person might save calculation effort by estimating mentally: 90 000 orbits in 15 years is 6000 orbits per year, 680 days is about 2 years, so the answer is about 12 000 orbits. In a real situation, appreciating the accuracy that is possible from the data, and that is required to sensibly answer the question, is critical.

Ferris Wheel (Q2), an interpreting task, is also about proportional reasoning, but part of the conceptual demand is visual and part dynamic – seeing the movement in a circle over time, including the direction. Again, there are various approaches. Probably the simplest is informal: to recognise that the numbers (40 minutes and 30 minutes) are simply related, then to see that in each 10-minute interval the wheel moves a quarter of a revolution, so point *S* is the answer. The situation is an interesting one even if it is not obvious why anyone would ask that question – perhaps when wondering where to look to wave to one's friends? One could generalise the argument, relating the angle travelled to the time taken.

The set of released tasks from PISA [2A] is a far richer resource for teaching mathematical literacy than these examples might suggest. It is a collection every teacher of students in the PISA age group should explore. However, the PISA task designers are constrained by the circumstances of an international comparison survey. Each task must take most students no more than a few minutes to complete, so that the complete survey can sample an adequate range of real-world context types, mathematical content, and mathematical processes. The task must also work in about 70 countries with their different cultures and languages – 'work' in the sense of reliably allowing students to understand the real-world situation, be likely to have learned the relevant mathematics by age 15, and then to show what they can do. A challenging brief indeed. The need for statistical robustness of the results provides further constraints. These and many other issues are discussed in depth in *Assessing Mathematical Literacy* (Stacey & Turner, 2015), their account of the design and delivery of the PISA 2012 survey. The multiple-choice formats often used in PISA, as in the original versions of these tasks, follow from these constraints. For classroom use, the tasks are usually best presented in an open format, requiring a constructed response and explanation.

Upgrade?	**Table Tennis Tournament**
My phone works fine but I've been offered an upgrade. How should I decide whether to take it? An *evaluate and improve* task	Three of us have agreed to organise a table tennis tournament for our club. How should we set about it? What information will we need? A *plan and organise* task

FIGURE 2.1.2 Challenging situations for investigation.

Questions that arise in everyday life are usually broader and messier than these tasks, so a mathematical literacy curriculum must include longer and less-structured tasks that demand more extended chains of reasoning than found in PISA, in typical classroom exercises, or in examinations. This means that students should be expected to reason their way through the multiple phases of modelling set out in Chapter 1: looking at what variables are important, what assumptions to make about the relationships between them, what data to collect to inform the analysis, and how to interpret and evaluate the results – and to do all this without being led to a particular solution by the teacher. Similarly, in a data-driven situation, students need to interpret it not just by examining the data given and its statistical properties, but also considering its import in the context. In both cases, this also means students having time to reflect on, discuss, and revise solutions. This is all required for the two authentic tasks in Figure 2.1.2. Students will need preparation to learn how to deal effectively with the demands of these very open, extended tasks.

Figure 2.1.3 gives two examples of PISA tasks that can be used to help students develop an investigative approach, understanding what is required and developing confidence. Few students solved the *Revolving Door* task under PISA test conditions, but it can work well in the classroom with teacher support through discussion. Although constrained, these tasks are more open and involve more extended thinking than is usual. This will at first be a challenge for students, and a challenge for their teachers to support them, so the examples that follow range from tasks only slightly longer and less structured than typical test questions through to projects that can fuel several lessons' work. The suggestion is to build into the curriculum a spectrum of tasks with gradually increasing demand.

2.2 Sample 5–10-minute tasks

The tasks in Figures 2.2.1 and 2.2.2 (overleaf) each take about 5–10 minutes of class time. *Which Car* is a PISA question where the main challenge is absorbing and reviewing fairly complex data and then relating it to the specified constraints.

Continent area

Estimate the area of Antarctica using the map scale

Show your working out and explain how you made your estimate. (You can draw over the map if it helps you with your estimation)

An interpret task

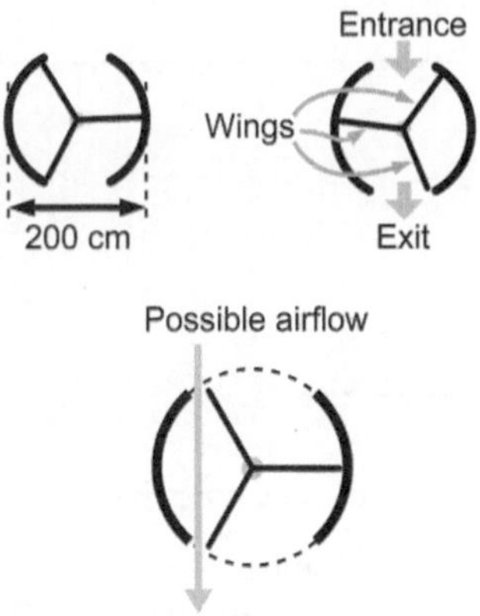

Revolving door

A revolving door includes three wings which rotate within a circular-shaped space. The inside diameter of this space is 2 metres (200 centimetres). The three door wings divide the space into three equal sectors. The diagrams show the door wings in three different positions viewed from the top.

Q2: The two door openings (the dotted arcs in the diagram) are the same size. If these openings are too wide, the revolving wings cannot provide a sealed space and air could then flow freely between the entrance and the exit, causing unwanted heat loss or gain. This is shown in the diagram opposite.

What is the maximum arc length in centimetres (cm) that each door opening can have, so that air never flows freely between the entrance and the exit?

FIGURE 2.1.3 Two richer tasks from PISA.

Source: PISA Released Mathematics Items [2A].

Which car?

Chris has just received her car driving licence and wants to buy her first car.

The table below shows the details of four cars she finds at a local car dealer.

Model:	Alpha	Bolte	Castel	Dezal
Year	2003	2000	2001	1999
Advertised price (zeds)	4800	4450	4250	3990
Distance travelled (kilometres)	105 000	115 000	128 000	109 000
Engine capacity (litres)	1.79	1.796	1.82	1.783

Chris wants a car that meets all of these conditions:
The distance travelled is not higher than 120 000 kilometres.
It was made in the year 2000 or a later year.
The advertised price is not higher than 4500 zeds.

Q1: Which car meets Chris's conditions?

FIGURE 2.2.1 An *evaluate and improve* task based on data.

Source: PISA Released Mathematics Items [2A].

Which Sport?

Which sport will produce a graph like this?

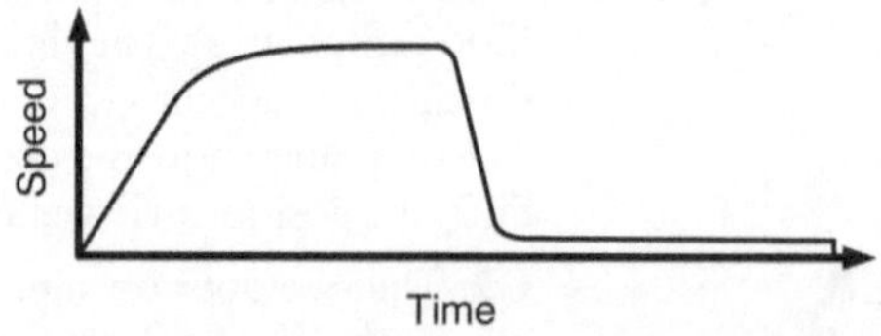

Choose the best answer from the following and explain exactly how it fits the graph.

Fishing
Pole Vaulting
100 metre Sprint
Sky Diving
Golf
Archery
Javelin Throwing
High Jumping
High Diving
Snooker
Drag Racing
Water Skiing

FIGURE 2.2.2 An *interpret* task.

Source: The Language of Functions and Graphs – Shell Centre for Mathematical Education [2B].

This is typical for tackling this kind of consumer choice decision. The straightforward strategies are to take the cars one by one checking the data against each constraint, or to take the conditions one by one and eliminate cars at each stage. Alpha is eliminated on price, Castel on distance, and Dezal on age. Bolte scrapes through on all three which, in classroom use, could stimulate a critical thinking discussion: "Is one year older serious when you could buy Dezal and save 460 zeds, about 10% of the price of the car?" It is worth noting that the engine capacity numbers are irrelevant to choosing on these criteria – sensible since they are closely comparable; identifying which data is significant and discarding redundant information is an essential part of mathematical literacy.

Which Sport? (Figure 2.2.2) is also about 'getting meaning out of data' – here a speed-time graph. The presentation invites the common 'picture-graph' confusion, arising when students do not think carefully about the variables on the axes. This will emerge and be sorted out in group and class discussion. This task is from *The Language of Functions and Graphs* [2B] for which the designer, Malcolm Swan, was awarded the prize for design excellence of the International Society for Design and Development in Education (ISDDE). 'LFG' is an excellent teaching resource on the interpretation and sketching of line graphs of real situations – a competence that is so important for getting meaning out of data of many kinds.

Asking for permission (Figure 2.2.3) – a *design and make* task – explores the other, human interaction, end of the demands of mathematical literacy. What information will the parent need to give informed permission for their child to go on

Asking for permission

Imagine that *your* class is going on a trip to *your* local swimming baths next Tuesday morning. One person has volunteered to write a letter to parents, asking for permission.

Make a list of all the important details that have been missed out of the letter.

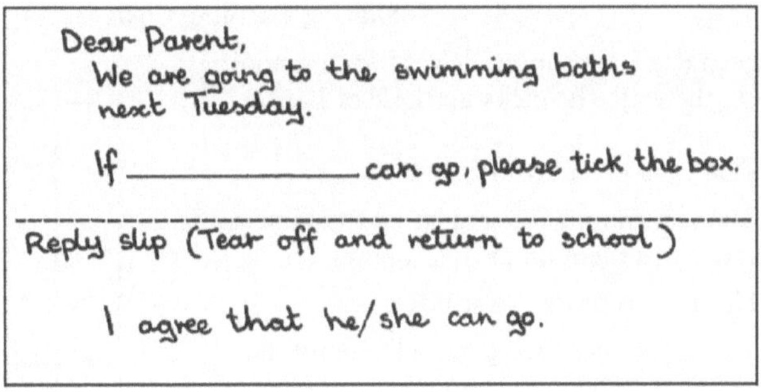

FIGURE 2.2.3 A *design and make* task.

Source: Plan a Trip – Shell Centre for Mathematical Education [2E].

this trip to the pool? How can it be presented in a convenient form that is likely to get a positive response. This task requires logical, rather than mathematical, thinking. It is included to show a short task that can scaffold students' thinking and build confidence to help them work an extended task, in this case the *Plan a Trip* module (see Figure 2.4.1) from the *Numeracy through Problem Solving* series.

Students need time to consider each task carefully, and to work through it; how much time depends on how far the teacher wants to take student explanations and discussion – and consider the extensions that may emerge. A core aspect of developing mathematical literacy in the classroom is tackling fewer, more substantial tasks and thinking through the meaning of the mathematics involved in greater depth. Slower can be quicker in learning.

2.3 Sample lesson-length tasks

Tasks that arise in everyday life often do not have 'right answers' – though they nearly always have wrong ones! The tasks in this section come from the 'Classroom Challenges' lessons of the Mathematics Assessment Project [2C]. These materials provide substantial support for teachers. Many of the lessons focus on concept development, designed to reinforce and deepen understanding of curriculum topics, but about a third are problem-focused and highly relevant to mathematical literacy. The project is described in Appendix 1 of this chapter.

Sharing gasoline costs (Figure 2.3.1) requires careful reasoning, with no right answer. Students find issues of fairness important and motivating.

Sharing Gasoline Costs

Each day Lara's mom drives her to school.
On the way, she picks up three of Lara's friends,
Chan, Jason and Marla.

Each afternoon, she returns by the same route and drops them off at their homes.

This map is drawn to scale.

It shows where each person lives and the route taken by Lara's mom.

At the end of a term, the four students agree to pay $300 in total towards the cost of the gasoline.

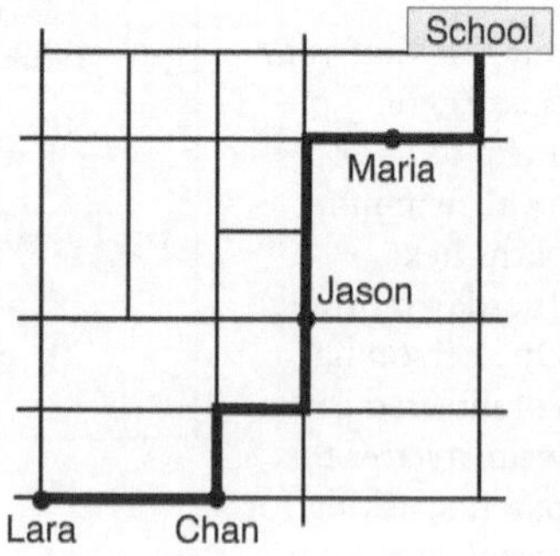

How much should each person pay?
Try to find the fairest possible method. Show all your work.

FIGURE 2.3.1 A middle school *evaluate and improve* task.

Source: Mathematics Assessment Project *Classroom Challenges* [2C].

This leads to discussions in some depth on how to define fairness in this context, each approach leading to a different algorithm to partition the cost. For example, some students will charge in proportion to distance travelled; others might charge less for blocks where there are more people in the car. Others point out different aspects of the context that can tip the balance for fairness – should savings from not catching the bus be considered? Sample student solutions for discussion are given in the online lesson resources. An *evaluate and improve* task where students compare various algorithms, this is also the focus of a professional development module on *Questioning and reasoning* from the *Bowland Maths* materials [2D].

Making fair divisions of cost is a rich topic for strengthening students' critical thinking about the appropriateness of a direct proportion model in different circumstances. As another example, how can two friends share the cost fairly, if they buy two dresses together, using an offer of "buy one dress, get 40% off a second (less expensive) dress". Several potential models come to mind: pay half the cost each, get half the saving each, divide the total cost in proportion to the original prices. Making the fair decision involves details of the social context as well as mathematical understanding of the consequences of different procedures.

Making matchsticks [2C] (Figure 2.3.2), though the context is not directly related to students' lives, epitomises estimation problems and the mathematical challenges involved. The core concept here is ratios of volumes. Key technical challenges are in handling big numbers, working to appropriate

Making Matchsticks

Matchsticks are $\frac{1}{10}$ inch by $\frac{1}{10}$ inch by 2 inches.

Matchsticks are often made from pine trees.

Estimate how many matchsticks can be made from this tree:

80 feet tall

2 feet diameter at the base.

You may find some of the information given on the formula sheet helpful.

Explain your work carefully, giving reasons for any choices you make.

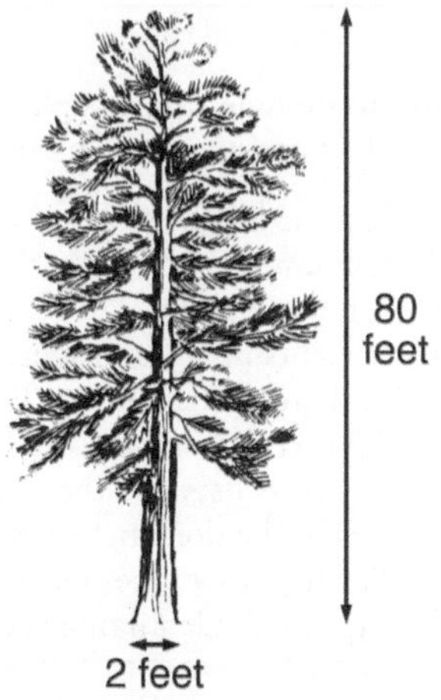

FIGURE 2.3.2 A high school task on volumes involving estimation.

Source: Mathematics Assessment Project *Classroom Challenges* [2C].

accuracy and dealing with the units correctly. Formulating the model involves recognising that this can only be a rough estimate, cutting out the complexities of the tree and choosing a simple shape for the volume calculation: cylinder or cone? A *model and explain* task, designed for the United States, the non-decimal 'customary units' still in use there increase the challenge in some ways but, because the factors 12 are explicit, errors made in the conversions of length and volume stand out.

2.4 Sample multi-lesson tasks

Finally in this chapter, Figure 2.4.1 outlines examples of two mathematical literacy tasks that are of project length. *Plan a Trip* is a *plan and organise* task from the UK Shell Centre *Numeracy through Problem Solving* project [2E]. *Similar Triangles (Ironing Table)* is an *explore and discover* task from the set of 16 substantial lessons for Years 7–10 from the special topic *Mechanical Linkages and Deductive Geometry* by the Australian reSolve Project [2F]. The following outlines of these two tasks aim to help teachers decide whether to investigate these rich sets of online teaching materials.

Numeracy through Problem Solving is a sequence of five modules that develop students' mathematical literacy. The five modules are *Design a Board Game*, *Produce a Quiz Show*, *Plan a Trip*, *Be a Paper Engineer*, and *Be a Shrewd Chooser* [2E]. They were designed by Malcolm Swan and the Shell Centre team and developed with typical British teachers and their mixed-ability classes. Designed for the age range 12–16, they have been used

Plan a Trip

In this module students plan and undertake a class trip out of school. This involves costings, scheduling, surveys, and everyday arithmetic.

- In a card game simulation, groups undertake and record imaginary trips, encounter problems and errors of judgment, then seek to correct them by better planning.
- Students in a group share ideas for possible places to go and produce a leaflet explaining these ideas. The class then work together to reach a decision on the best destination and look at possible means of transport.
- The class lists and then shares out and undertakes the preparatory tasks that need to be done before the trip can take place.
- The trip now takes place and, afterwards, the students reflect on what happened, identifying successes and failures.

FIGURE 2.4.1 A *plan and organise* mathematical literacy project.

Source: Numeracy through Problem Solving – Shell Centre for Mathematical Education [2E].

successfully with younger children and with adults. Each module is designed to take between 10 and 20 hours, usually over three weeks. The work is primarily guided by a student booklet, with the teacher playing a monitoring and 'consultant' role.

Each module works on a group-project basis and has four stages (exemplified in Figure 2.4.1, for planning a day trip out of school). In Stage 1 students explore the context by working on and evaluating exemplars provided. Stage 2 is about generating and sifting ideas, which are developed and implemented in detail in Stage 3. In Stage 4, each group evaluates the designs that the other groups have produced. These are the essential phases of most real-world problem-solving. A further example from this series, *Produce a Quiz Show*, is presented in Section 6.2. The Shell Centre Publications website gives further information on the modules and access to teaching materials.

Two kinds of assessment for these modules were provided. Formative assessment, built into the teaching materials, is designed to check that each student in a group understands all aspects of the work, not simply those for which they may have been responsible. A final examination at the end of the module assessed how well students could transfer what they learned to more or less closely related problem situations. Administered as formal examinations by the examination board, it showed one way that mathematical literacy can be assessed in a high-stakes accountability situation.

Mechanical Linkages and Deductive Geometry [2F], a *reSolve: Mathematics by Inquiry* special topic with four units aligned to the Years 7–10 curriculum, shows students how geometry makes many things work. It provides important applications of geometric theorems and stimulates the need for proof by asking students to be sure that the things will always operate as required. Mechanical linkages – sets of hinged rods – form the basis of many everyday objects such as folding umbrellas, car jacks, scissors lifts, toolbox lids, and pantographs. These objects work properly because of the geometry of triangles, quadrilaterals, and circles. The lessons offer rich potential for group exploration and justification. Students start by observing a real-life object, identifying what must occur. For example, a car jack must rise vertically. They then make a working model of the linkages from plastic strips or card and paper fasteners and later use pre-prepared dynamic geometry software programs to investigate the underlying geometry. The software allows students to observe more clearly what stays the same and what varies as the models operate. The lessons are focused on angles and lines (Year 7), quadrilaterals (Year 8), similar triangles (Year 9) with proof, and circle geometry for Year 10. Extensive resources for teaching, including model templates, teaching notes, and slides are provided (Figure 2.4.2).

Ironing Tables – from reSolve Mechanical Linkages (Similar Triangles unit)

When an ironing table with legs that pivot is raised or lowered, the top must always stay parallel to the floor. How does this happen? Students investigate the triangles formed by the pivoting legs, in particular investigating how different leg lengths and pivot positions ensure that similar or congruent triangles are formed. Three different designs can be considered, involving geometry of different complexity.

- Students examine an ironing table or similar small folding table and identify its important features, including size and shape for convenient ironing, need for stability, adjustable height, and neat folding away. They may notice some geometric features, such as the position of pivot point.
- Students create a physical model of the ironing table, observing more geometric properties, especially as it moves.
- Using the dynamic geometry program, they measure angles and lengths, and from this observe and later prove that two similar triangles are formed. These make equal angles, so that the table is always parallel to the floor.
- Students are challenged to design an adjustable table with unequal legs.

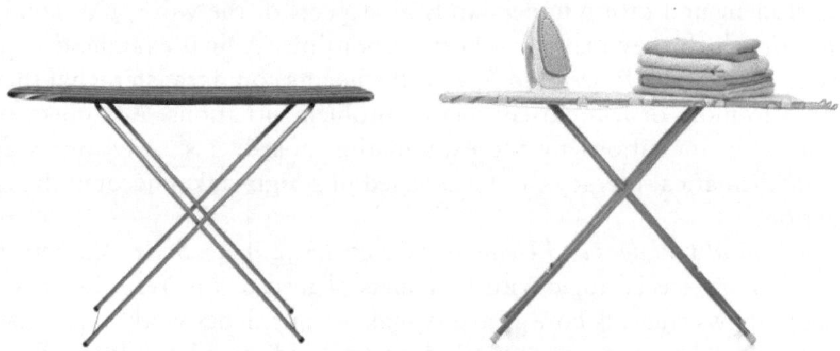

FIGURE 2.4.2 An *explore and discover* lesson involving movement of everyday objects.

Source: Mechanical Linkages and Deductive Geometry – reSolve [2F].

2.5 Task attributes

The examples of tasks in this chapter aim to illustrate something of the variety of 'know how' task types we will meet the chapters that follow. They give some idea of what mathematical literacy, in all its variety, is about – as well as pointing to some well-engineered resources for teachers to use in exploring mathematical literacy in the classroom. We now draw these examples together by listing six characteristics that will help in selecting a collection that represents a reasonably 'balanced diet' for developing the five components of the context-focused mathematics framework (Section 1.1) across a diverse range of contexts.

The type of task

Students should over time have experience of the six task types which began this chapter:

- Model and explain real-world phenomena.
- Plan and organise events.
- Design and make an artefact, or design a procedure.
- Explore situations and discover relationships.
- Interpret representations in words, numbers, graphs, or algebra to extract meaning.
- Evaluate and improve a solution to any of these.

The kind of context

PISA (see Stacey & Turner, 2015) divided the range of situations a task may involve into four broad context categories:

- personal tasks of immediate relevance to the individual,
- societal tasks supporting informed citizenship,
- occupational tasks relevant to future employment, and
- scientific tasks (including science, engineering, economics, and pure mathematics).

PISA includes situations from all of these categories to improve the validity of its measures. Similarly, teachers can include all of these categories.

Most of the situations addressed in this book address personal and societal contexts and concerns – the heart of mathematical literacy for most people – though preparing for the other two kinds is very important. Especially for young adults in vocational education, a focus on specific occupational tasks can stimulate motivation for learning mathematics that they have not experienced before. Mathematical literacy for scientific issues also has a workforce benefit and often clear societal and personal implications – implementing health advice sensibly, informing voting, and even knowing about the value of using logarithmic scales to report sound and earthquake intensity. Governments invest in mathematics teaching because of the benefits of having a generally mathematically literate workforce. There are mathematical literacy demands specific to each job or profession, some basic and some at a very high level.

Authenticity

Traditional school mathematics contains many 'real-world' tasks, like those in Figure 1.7.1, where the context is not intended to be seriously examined by the students. But because mathematical literacy is characterised by taking the context seriously, tasks used have a responsibility to be as authentic as possible, given classroom constraints. This means they all need to be fairly realistic with reasonable data, a focus on the big factors influencing the situation, expecting sensible solutions, and be clear whether the scenario is true or invented. Answers need to be given to a sensible level of precision, reflecting the precision in the data, assumptions made, and the demands of the context. Tasks that foster mathematical literacy most strongly use contexts that have an influence on the solution path. Of course, this does not exclude the occasional fantasy situation or a classic problem that beautifully, if implausibly, illustrates a mathematical point (such as the grains of wheat/rice on a chessboard task [2G]. A mathematically literate person will appreciate how extremely implausible the chessboard scenario is.)

Levels of expertise – 'novice', 'apprentice', and 'expert' tasks

A problem can be presented at various levels of sophistication to suit the level of expertise of the solver. An 'expert' task is one presented as it might arise in life, providing little or no guidance for how the problem is to be tackled, while an 'apprentice' version will break the problem down into several manageable steps. A 'novice' task will either be an isolated exercise focused on a specific mathematical procedure, or a collection of such tasks loosely grouped together under a context (such as series of calculations that might be encountered 'at the shops'). Developing mathematical literacy is about improving 'expertise', as revealed in responses to expert tasks; the other two types are important along the way.

Task difficulty

Traditionally, the difficulty of a mathematics task is judged primarily on the technical level of the mathematics involved (e.g. on a scale from addition to differential equations). However, the difficulty for students is also to affected by the complexity and familiarity of the task situation, and the autonomy expected of them in tackling the task. Unfamiliar tasks will be more difficult than well-practised exercises, even where the underlying mathematics is the same, as will tasks with more variables or more data, and those where the student is expected to work out on their own how to approach the task, and how to organise their work. So to keep tasks within the capacity of students, less familiar and more complex tasks must be technically simpler – hence Lynn Steen's description of mathematical literacy as "the sophisticated use of elementary mathematics" (Steen & Forman, 2000).

Openness

Many life-related problems don't have a single formulation or a well-defined correct answer, but allow for a diverse range of useful responses, possibly at different levels of sophistication. Collections of tasks designed for mathematical literacy should include such open tasks and encourage critical thinking to evaluate solutions from multiple viewpoints.

A more detailed discussion of the aspects of task design that foster mathematical literacy can be found in Chapter 14.

Appendix 1. The mathematics assessment project's Classroom Challenges

The Classroom Challenges [2C] were developed between 2010 and 2015 as part of a program to support the introduction of the US Common Core State Standards in Mathematics. The aim was to produce teaching materials which would encourage the formative assessment practices in Mathematics (Burkhardt & Schoenfeld, 2019) that had already proven highly successful (Black & Wiliam, 1998). This model of lesson enrichment also reflects Japanese approaches. About a third of these formative assessment lessons focus on non-routine problem-solving, some of these in life-related contexts.

The MAP team developed some design tactics that teachers may find useful in promoting mathematical literacy, and elsewhere where a more robust understanding of mathematics is desired. The top-level strategy is to look in greater depth than usual at the task – of course, choosing tasks that are rich enough to warrant that attention. To this end, the design of the problem-solving lessons employs two particular design tactics that are worth noting:

- A *Common Issues* table that lists difficulties that students showed in trials of the materials, each linked to suggestions for non-directive interventions (mostly questions) that a teacher might use to help the student move their thinking forward and surface any underlying misconceptions that may be blocking progress.
- *Comparing student work* where students are asked to compare carefully constructed examples of 'student work' that use different methods, often including numerical, graphical, and algebraic solutions. Besides making the point that there are usually multiple valid ways to solve a problem, this helps ensure that key elements of the curriculum are not lost and helps students recognise why more sophisticated methods can be more powerful.

Each *Classroom challenge* begins with a 15–20-minute task that students tackle individually in a prior lesson; this gives the teacher a chance to review the responses, looking for different approaches and common issues. (*Not* scoring, which is as counterproductive for formative assessment as it is time-consuming and it distracts from misunderstandings.)

Developing your solution. The main lesson begins with any general comments the teacher chooses to make, then 10 minutes or so for each student to review and reconsider their 'solution', before moving to work with one or two others for 20 minutes or so to produce a joint solution (felt-tipped pens on poster-size paper allows the teacher to observe without 'entering the group'). The teacher observes, noting different approaches, only intervening when a group is stuck – by choosing questions, perhaps from the Common Issues table, that help the group think of alternative approaches, avoiding any specific solution path. Groups then share their posters, encouraged to think about questions as follows:

- Did they choose a good method for representing the situation?
- Did they make sensible assumptions?
- Is the reasoning correct? Are the calculations accurate?
- Are the conclusions sensible?
- Was the reasoning easy to understand and follow?

Collaborative analysis of the sample responses is the next main activity (in the next class period if these are short). Each group discusses the three or four sample responses provided, asking these questions and developing written comments – assuming a 'teacher role'. This activity gives students the opportunity to see further approaches, and representations, and discuss the assumptions made in each case. These are then shared in a *whole-class discussion* of the relative merits of various approaches. Finally, each group returns to review its own solution and discuss possible changes it would make.

This outline will, as always, be brought to life by specific examples – working through the lesson plan of one of these MAP lessons (see link [2C]), for example:

Rolling cups – about the design of paper cups in the shape of a truncated cone

or

Having kittens – modelling the number of descendants of an un-neutered female cat

or

Muddying the waters – modelling the pollution of a lake as circumstances change

The Math Assessment Project materials have seen considerable success, with millions of downloads of individual *Classroom Challenges* modules. Independent evidence shows substantial learning gains arising from their use as part of a professional development program (CRESST, Research For Action, 2015).

Acknowledgements

Extracts from *The Language of Functions and Graphs* (Figure 2.2.2), *Numeracy through Problem Solving, Plan a Trip* (Figures 2.2.3 and 2.4.1) are free for education use, and the Mathematics Assessment Project *Classroom Challenges* (Figures 2.3.1 and 2.3.2) are available under the Creative Commons CC BY-NC-ND 3.0 licence. Both of these appear here courtesy of the Bell Burkhardt Daro Shell Centre Trust.

Several of the tasks referenced in this chapter are taken from the OECD's *Program for International Student Assessment* (PISA). These are adaptations of an original work by the OECD. The opinions expressed and arguments employed are the sole responsibility of the author or authors of the adaptations and should not be reported as representing the official views of the OECD or of its member countries. The adapted PISA items are licensed under the Creative Commons Attribution-NonCommercial-ShareAlike 3.0 IGO (CC BY-NC-SA 3.0 IGO) licence. The specific PISA items used in this chapter come from the Released Mathematics Items for 2006 and 2012:

Figure 2.1.1 *Space flight* – 2006 item M543Q01
Ferris wheel (Q2) – 2012 item PM934Q02
Figure 2.1.3 *Continent area* – 2006 item P01480
Revolving door – 2012 item PM995 (introduction and Q2)
Figure 2.2.1 Which car? – 2012 item M985Q01

Figure 2.4.2 is taken from *reSolve: Mathematics by Inquiry* © Australian Academy of Science, available under Creative Commons CC-BY-NC-SA licence.

See the Links section for online access to all of the above materials.

References

Black, P., & Wiliam, D. (1998). *Inside the black box: Raising standards through classroom assessment*. King's College.

Burkhardt, H., & Schoenfeld, A. H. (2019). Formative assessment in mathematics. In R. Bennett, G. Cizek, & H. Andrade (Eds.), *Handbook of formative assessment in the disciplines* (pp. 35–67). Routledge.

CRESST, Research for Action. (2015). *MDC's Influence on Teaching and Learning. Research for Action*. Retrieved from https://www.researchforaction.org/research-resources/mdcs-influence-on-teaching-and-learning/ Accessed July 1, 2023.

Stacey, K., & Turner, R. (Eds.) (2015). *Assessing mathematical literacy*. Springer.

Steen, L. A., & Forman, S. L. (2000). Making authentic mathematics work for all students. In A. Bessot & J. Ridgway (Eds.), *Education for mathematics in the workplace* (pp. 115–126). Springer Netherlands.

Links to useful material

To visit any of these links, scan this QR code or visit ltml.mathlit.org – append the link code to go directly to the entry – e.g. ltml.mathlit.org/2A

The original source links are given below for attribution purposes.

[2A] *PISA Released Mathematics Items 2012 and 2006* – OECD
https://www.oecd.org/pisa/
2006: https://www.oecd.org/pisa/testquestions-pisa2006.htm
2012: https://www.oecd.org/pisa/pisaproducts/pisa2012-2006-rel-items-maths-ENG.pdf

[2B] *The Language of Functions and Graphs* – Shell Centre for Mathematical Education
https://www.mathshell.com/materials.php?item=lfg&series=tss

[2C] *Classroom Challenges* – Maths Assessment Project
Sharing gas: https://www.map.mathshell.org/lessons.php?unit=6200&collection=8
Making matchsticks: https://www.map.mathshell.org/lessons.php?unit=8300&collection=8
Rolling cups: https://www.map.mathshell.org/lessons.php?unit=9300&collection=8
Having kittens: https://www.map.mathshell.org/lessons.php?unit=9100&collection=8
Muddying the Waters: https://www.map.mathshell.org/lessons.php?unit=9400&collection=8

[2D] *Questioning and Reasoning* – Bowland Maths
https://www.bowlandmaths.org.uk/pd/pd_05.html

[2E] *Plan a Trip* – Shell Centre for Mathematical Education
https://www.mathshell.com/materials.php?&series=numeracy

[2F] *Mechanical Linkages and Deductive Geometry* – reSolve
https://www.resolve.edu.au/mechanical-linkages

[2G] *Wheat and Chessboard Problem*
https://ltml.mathlit.org/2G

3

TEACHING FOR MATHEMATICAL LITERACY

This chapter addresses the teaching of mathematical literacy – helping students gain more power over the real-world situations they will encounter in their present and future lives. Education in mathematics can empower students to think more critically and constructively about their world.

3.1 The challenge

Mathematics, here including statistics, is a compulsory subject at school, studied for multiple hours a week over many years. Why? Largely because society as a whole appreciates that mathematics can make a unique contribution to understanding practical situations and supporting good decision-making in both personal and working lives. Why does mathematics have so much more curriculum time than, say, music?

Having an adequate level of mathematical literacy is important for the life chances of individuals. For the same reason, it is increasingly seen as having national benefit in our interlinked, competitive, and digital world. Of course, mathematics is not only taught for its utility: like music, it is also taught as a significant part of cultural heritage with its elegant structures, intriguing problem-solving opportunities, amazing theorems, and the special contribution it makes to logical thinking.

Both of these sides of teaching mathematics are important, but mathematical literacy is the essential goal for *every* student. If students leave school without an understanding of how they can use the mathematics they have learned, a general appreciation of how others use it, *and* a propensity (productive disposition, critical thinking) for doing so when faced with challenges, then their mathematics education has not been really successful.

DOI: 10.4324/9781003303503-4

This is increasingly recognised by countries around the world. It is a major part of one of the United Nations Sustainable Development Goals for 2030 (n.d.), where Target 4.6 is as follows:

> By 2030, ensure that all youth and a substantial proportion of adults, both men and women, achieve literacy and numeracy.

This represents a challenge for both teaching and teacher education in mathematics (Burkhardt, 2021; Madison & Steen, 2008). This chapter describes ways in which this challenge can be more effectively met. There are three main themes. The first is that developing mathematical literacy can be a part of the majority of mathematics lessons from the beginning to the end of education. Mathematical literacy can be fostered whenever teachers take the time to link new mathematics to life-related situations. This involves teachers taking the trouble to include problems and exercises with some degree of authenticity, ensuring their students expect the mathematics to make sense and, equally, that students think critically, *using their common sense* in discussing how realistic the mathematical solutions are. However, it is also essential that some lessons specifically target mathematical literacy, because of the *different priorities*, summarized in Figure 1.4.3, when the focus is on understanding the real-world situation. Then the mathematics, usually already well understood, provides tools that help to illuminate the situation. This book gives many examples of what these lessons might be.

The second theme, which will be addressed throughout this book, is that mathematical literacy changes as the world changes, and change is especially fast now as we are living in the digital revolution. The fluent pen-and-paper mathematical skills that used to be essential for everyday life, or to get a well-paying job, are rarely important now. Instead, a mathematically literate person is adept at using the mathematical tools that they have at hand, from a simple calculator on a smartphone to a spreadsheet on a work or home computer to programming very specialised software. School curricula around the world are slowly adapting to this new situation but continuing adaptation will be needed for decades to come. Technology advances on a timescale much shorter than curricula can change. An important part of developing mathematical literacy is teaching students to use commonly available mathematical tools well. In the tasks presented throughout the book, we assume students have access to appropriate digital tools.

The third theme underlying the chapter is the dependence of mathematical literacy on particular emphases in learning mathematics. Although every piece of mathematical knowledge is potentially useful to illuminate some life-related situation, some content is used very frequently, so deserves special attention. Using mathematics also requires more reliable mathematics. Beyond the classroom, there are likely to be many steps in solving real-world

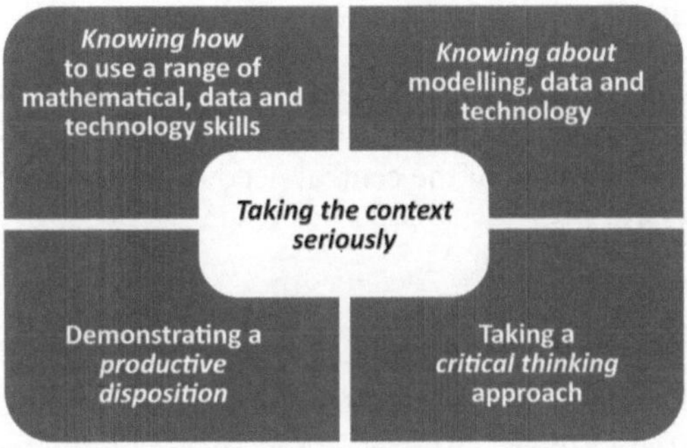

FIGURE 3.1.1 Context-focused mathematics framework.

Source: See Section 1.1.

problems, so students need *robust* skills and some autonomy in constructing and explaining chains of reasoning, using problem-solving and modelling strategies, as well as a capacity for 'self-monitoring'. These attributes are summarised for mathematical literacy in the five key components of our context-focused mathematics framework that we set out in Chapter 1 (Figure 3.1.1).

The implications for teaching will provide a connecting thread throughout this chapter. Section 3.2 discusses how teaching for mathematical literacy fits within the broader pedagogy of teaching mathematics, pointing to variations in teaching strategies that support mathematical literacy. In Sections 3.3 and 3.4 we look at the main topics that are taught in school curricula and highlight some of the aspects of each topic that are especially important for mathematical literacy. The idea is that developing mathematical literacy can be part of almost every mathematics lesson. Section 3.5 discusses the value of interdisciplinary activities, which give opportunities for students to engage deeply with the context. Here we suggest that mathematics teachers sometimes work with teachers of other subjects so that the life-related problems that are being studied can be treated with both mathematical depth and rich understanding of the context. Section 3.6 looks at the 'critical orientation' that is part of mathematical literacy as a complex competency; in particular, competency in examining evidence and creating and communicating arguments and conclusions is recognised as part of the twenty-first-century skills that are fundamental goals of education. Section 3.7 discusses what all mathematically literate students need to know about mathematical models – only a few people will ever create significant mathematical models, but everyone uses them, directly or indirectly, and needs the capacity to both appreciate and critique them. Finally in Section 3.8 we discuss the professional development for teachers that will help them help their students learn to become mathematically literate.

3.2 Pedagogical variations for mathematical literacy

The fundamentals of teaching for mathematical literacy are no different from those for teaching mathematics, and indeed other disciplines. But some differences in emphasis arise from the central need to understand the real-world phenomena involved.

The TRU dimensions

The essentials of fine mathematics teaching have been extensively discussed and researched over centuries – and increasingly probed in a wide range of research programmes. Here we shall use the distillation of best practice set out in the *Teaching for Robust Understanding* dimensions, developed by Alan Schoenfeld and teams of collaborators [3A]. TRU asserts that the quality of a learning environment depends on *the extent to which it provides opportunities for students* along the five dimensions shown in Figure 3.2.1.

How do these five dimensions relate to the key components of mathematical literacy set out in the context-focused mathematics framework of Figure 3.1.1? TRU dimensions 1, 2, and 5 – content, cognitive demand, and

1	Content	The extent to which classroom activities provide opportunities for students to become knowledgeable, flexible, and resourceful mathematical thinkers
2	Cognitive demand	The extent to which students have opportunities to grapple with and make sense of important ideas and their use through *productive struggle*
3	Equity	The extent to which classroom activity structures invite and support the active engagement of *all* students
4	Agency, ownership, and identity	The extent to which students have opportunities to feel that the insights they develop are theirs and their fellow students', rather than just the teacher's. This contributes to their development of willingness to engage and of positive identities as thinkers and learners
5	Formative assessment	The extent to which classroom activities and assessment elicit student thinking and how subsequent teaching responds to those ideas, building on productive beginnings and addressing emerging misunderstandings

FIGURE 3.2.1 The five dimensions of powerful classrooms.

Source: Adapted from Schoenfeld et al. 2016 [3A].

formative assessment – are central to both knowing how to use mathematical, data, and technology skills and knowing about modelling, data, and technology – two of the five components of the framework.

TRU dimension 1 – content

Mathematical literacy draws on robust and flexible knowledge of mathematical content, built in classrooms that strongly exhibit these three TRU dimensions. As noted in earlier chapters, sufficient 'know how' for mathematical literacy means rather different things in different life-related contexts, and for people at different levels of mathematical sophistication, but everyone benefits from stronger understanding of mathematical and data content that they have learned. Similarly, everyone benefits from having a good understanding of how to use technology in solving mathematical problems. In Sections 3.3 and 3.4, we highlight some of the topics that are especially important for the *know how* component of mathematical literacy. In Section 3.7, we address a key aspect of the *know about* component (knowing about modelling), with many examples in the later chapters.

Developing the *know how* to use mathematical modelling and the *know about* to appreciate its use in society needs specific teaching. To build deeper understanding of mathematics, the situations addressed need to include both models that describe standard applications of mathematics and a variety of non-routine real-world situations suitable for active modelling by the students, some of which were described in Chapter 2. Life-related situations are often messy; sorting out the variables that are important and identifying likely mathematical relationships between them is at the heart of mathematical literacy.

TRU dimension 2 – cognitive demand

Interesting life-related situations are often 'open tasks' where students can be encouraged to take further the depth and extent of their analysis – both the mathematical techniques used and the variables and data included. Taking the context seriously creates a domain where cognitive demand at a level which is conducive to productive struggle can arise naturally, as in many of the task examples in Chapter 2. The need for critical thinking about methods and results as part of mathematical literacy also raises the level of cognitive demand.

Of course, in classrooms, cognitive demand needs to be adjusted to a workable level. Because of the strategic demands of model formulation, the mathematical concepts and techniques that students can effectively deploy in active modelling will be limited to those which they have *thoroughly absorbed, and connected* to other aspects of mathematics and a range of applications – usually topics that were first taught a few years earlier. This 'few year gap' between imitative and autonomous deployment of mathematical skills needs to be recognised in selecting situations to analyse; complex non-routine tasks requiring recently taught mathematics will prove too difficult for most students. Again, the core of mathematical literacy for most people is Steen's (2002) "sophisticated use of elementary mathematics".

TRU dimension 5 – formative assessment for learning

This dimension emphasises that powerful instruction 'meets students where they are' and gives them opportunities to deepen their understandings and, through discussion in groups and with the class, 'debug' their errors and misconceptions. With or without formal assessment, this dimension points to the need for teachers to take time to examine how their students think about mathematics in real-world contexts.

TRU dimensions 3 and 4 – equity, agency, ownership, and identity

These TRU dimensions are especially relevant to developing the context-focused mathematical framework's broader components of *productive disposition* and *critical thinking*. Classrooms in which a small number of students get most of the 'air time' are not equitable, no matter how rich the content; all students need to be involved in meaningful ways. Here open tasks can offer different 'entry levels' and 'end points' to a problem. Students can begin work using mathematical constructs with which they are comfortable and then extend or refine solutions through group and individual work. *Note of caution:* To foster a student's agency, ownership, and identity as someone who can tackle problems successfully, pointing out potential use of more advanced mathematics in problem contexts should be retrospective, not undermining the value of solutions that used simpler mathematics.

The essential role of the life-related context in mathematical literacy provides opportunities for students to bring their own specialised out-of-school knowledge from hobbies and passions into a solution. Drawing on students' special interests can help them become more engaged with mathematics because they see its relevance to topics which relate to who they are or are likely to become. Even those who are usually unable to contribute much to conversations in mathematics can make contributions, build on others' ideas, and have others build on theirs. This experience helps to develop a *productive disposition*. Making comparisons between different people's approaches and evaluating mathematical insights against practical knowledge of the context encourages *critical thinking*.

An approach where mathematical literacy problems put the student in the role of a 'consultant', investigating the situation for a specific 'client-audience', is powerful in developing agency and a *productive disposition*. It has the additional value of requiring explanations of the analysis and of encouraging

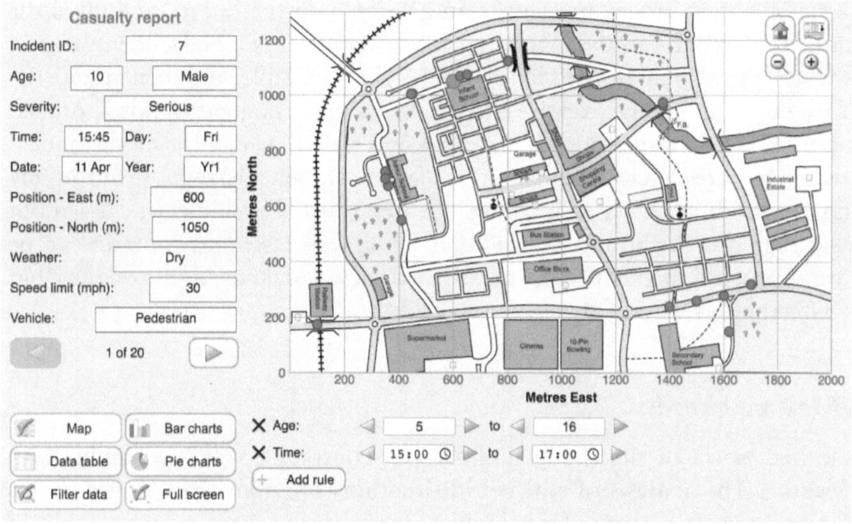

FIGURE 3.2.2 Reducing road accidents – exploring the data.

Source: Bowland Maths [3B].

more thorough investigation. A consultant knows that a client needs a full analysis and explanation. *Reducing Road Accidents* from Bowland Maths [3B] is an example of a lesson that exploits this: students are tasked with preparing a road safety plan for a small town. They are given a budget, a price list of various schemes such as speed humps, pedestrian crossings, cycle lanes, and education campaigns. They explore an interactive map-based database of past accidents (Figure 3.2.2) to collect evidence. The series of lessons culminates with students presenting their proposals to the class.

Some research (see e.g. Phillips et al., 1988) has shown that this kind of *role shifting* is an effective way to get students to assume more autonomy and responsibility for their work. Often it may simply be asking students to adopt roles that are normally their teacher's, such as explaining reasoning or asking questions – questions that demand reasoned explanations, not just answers.

Changing the classroom contract

Taking the context seriously, our central component of mathematical literacy presents some challenges for teaching and requires approaches that are new in some classrooms. Often this implies a change in the 'classroom contract' (Brousseau, 1997). This refers to the usually implicit understanding between teacher and students as to what each will do, and the roles each will play. For example, in the traditional '3X' or 'pitcher' model of 'instruction', the teacher's job is seen as filling the student up with knowledge – the teacher *eXplains* a new procedure, works an *eXample* on the board, and gives the students a set of *eXercises* involving closely similar tasks. The students expect to watch the explanation, to follow the worked example, and then imitate the procedure in doing the exercises. However valid the content is, this approach is not strong on dimensions 2, 3, and 4 of TRU! Teaching for mathematical literacy requires a 'classroom contract' where students have more autonomy and responsibility for their learning and are expected to think critically about what they do – a shifting of roles. Changing the classroom contract can be hard – proceed in small steps, persist with new demands, and explain what is different and why.

Reliable mathematics

Another aspect of the changed classroom contract involves reliability and checking. The analysis of real-world situations often involves longer chains of reasoning than most of the highly focused tasks that students tackle. If any link in the chain is broken, through a procedural error, for example, the subsequent reasoning and the conclusions are probably incorrect. In a set of short exercises, 80% correct may be seen as "not bad", but after four successive steps with a random 20% error rate, you are more likely to be wrong than right. More practice on short exercises is not the answer; instead, teachers can raise students' awareness of 'checking tactics'. In increasing order of the work likely to be involved, they are as follows:

- *Sense making:* Ask if the result makes sense in the context. If the result (whether final or intermediate) seems surprising, then it is either very interesting or (alas, more often) there was a mistake somewhere. In either case, check it.
- *Comparing different representations of the results:* This can be a big help in sense-making – graphs of various kinds may bring out unexpected features.
- *Parallel calculation:* Here, several students calculate separately, the results are compared and, if there are discrepancies, the 'bug' is hunted down.
- *Independent routes:* Seeking to 'do it a different way' is a very powerful checking method. Finding a different way almost always gives further insight into the context. Richard Feynman made this point to one of us:

  "If you understand a result one way and it seems interesting, it's worth pursuing. If you understand it two ways, it may well be right. If you understand it three ways, it almost certainly is". – Richard Feynman, as related by Hugh Burkhardt.

Of course, these checking methods only make the path from assumptions to consequences more reliable; if the assumptions are wrong, at least you know where to seek improvement.

A product

Creating a finished product from a substantial investigation into a real-world situation is very valuable. It creates a focus for multi-step investigations and is a concrete embodiment of a significant achievement by the student(s), reinforcing agency. The finished product may be a report, a presentation, or an artefact. Displaying work facilitates cognitive and meta-cognitive comparisons with the work of others, which is itself valuable for learning. One reasonably time-efficient way to organise this is a 'gallery walk', where students view each other's displayed work offering peer-review and constructive feedback focused on specific features to help each revise their product, perhaps leaving comments with sticky notes. The culmination of the *Reducing Road Accidents* lessons (Figure 3.2.2) was group presentations of their accident prevention plans to the class, who represented the town council. The presentations are supported by a poster or slideshow (examples are in the assessment guide [3B]). Having a product is a pervasive and powerful feature of Japanese mathematics classrooms, for example, which emphasize the effectiveness for learning of doing fewer activities in greater depth.

3.3 Orienting mathematical concepts and skills for mathematical literacy

This section outlines some concrete suggestions for how teaching 'normal' mathematics content can be oriented to the teaching of mathematical literacy. As we worked with the many mathematical literacy tasks that we considered in preparing this book, we created the list below to summarise the concepts, skills, and understandings that were most prominent. With just a little change, these topics can be reoriented to better foster mathematical literacy. Of course, this 'little adaptation' of content is not intended to be taught in an abstract way divorced from the real world.

A mathematical literacy approach to number will emphasise the following:

- Using mathematical tools, especially calculators (on any device) and spreadsheets.
- Quick estimation of the expected results of calculation.
- Calculating reliably (by whatever method) and checking results are sensible.
- Very large and very small numbers – their size and how to write them.
- Understanding and calculating with positive and negative powers of 10.

A mathematical literacy approach to measurement will emphasise the following:

- The size of units in relation to real-world phenomena. (How fast does 1 metre per second feel? How many Joules, or kilowatt-hours, to boil a litre of water?)

- Estimation of quantities, and knowing the typical measurements of some common objects to use as benchmarks.
- Units for very large and very small quantities (e.g. terabytes of data, petajoules of energy, nanometres) and their relationship to powers of 10 and 1,000.
- Reporting results of calculation with sensible precision.

A mathematical literacy approach to proportional reasoning will emphasise the following:

- All aspects of proportional reasoning and linking them (percents, ratios, rates, etc.).
- Dealing with rates in less usual units (e.g. number of deaths per 100 000).
- Combining several rates (e.g. finding number of drops per minute to set an intravenous drip using volume and time required and number of drops per litre).

A mathematical literacy approach to statistics will emphasise the following:

- Four stages: reading the data, reading between the data, reading beyond the data, and reading behind the data (Shaughnessy, 2007).
- Using technology (especially spreadsheet-like tools) to organise data sets, create data displays, and do calculations.
- Increasing the range of data displays that students encounter in line with changing practice (e.g. animated graphs).

A mathematical literacy approach to probability will emphasise the following:

- Including discussion of risk to supplement work on chance.
- Linking mathematical with common language and practices (e.g. betting odds).
- Visualising very small probabilities.

A mathematical literacy approach to geometry, spatial reasoning, and location will emphasise the following:

- Identifying geometry in the world around and appreciating that geometric features can make real objects work (e.g. a scissors lift keeps sections parallel, angles equal, and movement perpendicular to the ground).
- Using paper and digital maps and navigation tools.
- Visualising, interpreting, and making drawings and objects in three dimensions.
- Specific links to hobbies or vocational interests of students.

A mathematical literacy approach to algebra will emphasise the following:

- Reading and using formulas.
- Ideas of independent and dependent variables.
- Behaviour of basic types of functions (linear, exponential, inverse proportion).
- That linear functions describe constant rate of change and exponential functions describe a constant percentage rate of change (constant addition or subtraction versus constant multiplication or division at each step).

Underlying these recommendations is an expectation that students learn to use digital tools appropriately (see also Chapter 10) and that many of the learning activities involve authentic contexts.

3.4 Tweaking mathematics lessons towards mathematical literacy

In this section we look at three examples of how mathematics teaching practices can move more towards teaching for mathematical literacy. The first is the adaptation of word problems, so that students can experience the framework element of 'taking the context seriously'. Next, we look at how estimation (highlighted as an important emphasis under both number and measurement in Section 3.3) can be included in almost any mathematics lesson. The third example is about data displays, illustrating how curriculum needs to adapt to changing mathematical practices in society.

Word problems to mathematical literacy problems

What do we mean by enriching tasks to make the link to mathematical literacy stronger? Here is a typical word problem on exponential decay.

> The value of a car decreases by 15% a year. Anay buys a car for $10,000. How much will it be worth after 6 years?

The solution expected is to identify and use the formula, $FV = PV(1 - d)^n$, where FV = future value, PV = present value, d = rate of decay per period, and n = number of periods. In terms of our modelling process (Figure 1.3.1), the model has already been formulated (exponential decay at a rate of 15% a year) and all that is left to the student is the 'solve' stage – with minimal 'interpretation' and then 'reporting' the answer of $3771.495 to a specified accuracy and remembering to include the dollar sign. Students encounter many 'word problems' of this type where the task is essentially to spot the right formula and calculate the answer. They are intended to strengthen skills in identifying the formula, selecting the right data, and calculating correctly.

They do add to students' appreciation of how formulas are useful in the real world, in this case flagging that exponential decay is important for predicting depreciation of assets. Stillman (1998) and also Brown (2019) label such a real-world context as a 'border' context, which can almost be ignored by the solver except to choose the formula giving loss of value rather than gain.

How could we enrich this task to better foster goals of mathematical literacy? One change is to ask students to investigate the assumption of exponential decay (the missing formulation and evaluation steps). Another is to build in some practical decision-making (bringing in interpretation and reporting). The following version does both. In Stillman's terms, this is a 'tapestry' context, where the real-world task and the mathematics are woven together.

Anay buys a car for $10,000. Her friend Sam says a car loses 15% of its value every year.

a Find a way to check if this model is realistic.
b According to the model, how much is the car worth after 1 year?
c Calculate the modelled value of the car over time – you might use algebra or a spreadsheet.
d After owning the car for 4 years, it breaks down. Anay finds out that she will need to replace the clutch to be able to drive the car again. How could she find out if that is a good investment?

The curriculum probably requires students to practise the algebraic solution using the exponential function, but the second version permits the student to use alternative methods (e.g. spreadsheet) and provides some support in part (b). The second version also requires students to search for their own data on used car prices and the cost to replace the clutch. All of this takes longer than the first version (although data could be supplied when time is short), but including a certain number of 'tapestry' problems amongst with 'border' problems strengthens a focus on mathematical literacy.

The original task has probably been formulated in that way to mimic the assessment requirements: its focus on a single technical skill (solve using the right formula); without access to everyday computing tools; a self-contained question for an individual with no opportunity for research or discussion; quick to complete – all leading to a single, easily scored right-or-wrong answer. Importantly, there is certainly no need for *teaching* to be hamstrung by those requirements by using the same style of task exclusively.

Estimation and appropriate accuracy in every lesson

This section examines estimation, to illustrate how the somewhat different requirements of mathematical literacy can fit into and enhance the normal curriculum. Accuracy and precision are seen by many as a hallmark of mathematics. Who else but mathematicians want to calculate millions of digits of pi? But for mathematical literacy, quick estimation is often more important. Estimation and selecting appropriate accuracy to report results can play a role in many mathematics lessons.

Estimation is a multifaceted skill, relying on teaching in both number and measurement. It is essential in all applications of any area of mathematics, from quickly working out how much money to hand over to pay in a shop and to check your change (estimation as rough calculation), to checking numerical work (is my answer to these precise calculations reasonable), to buying enough lace to trim a dress (estimating a quantity), to providing a quote to renovate a bathroom (combining many reasonable assumptions). When a precise answer has been calculated (e.g. by hand or using a calculator), decisions also need to be made about the sensible accuracy with which to report the result. Is $3771.495 a reasonable answer for the value of Anay's car after 6 years? What is sensible in that context?

Fortunately, nearly every lesson brings up some opportunities to estimate. Quick checking of calculations is always important – in class as in real life. This involves number-sense (is it reasonable that 5/13 is 3.6 – of course not, I accidentally divided 13 by 5 instead) and can be a creative activity, because there are no prescribed ways to estimate. Instead, students can bring whatever they know to estimate answers. They might check against physical experience: is it reasonable that the third side of a triangle with two sides equal to 1 metre is 4 metres? Of course not, it is always shorter to walk directly from A to C rather than going via B, so I expect an answer between 1 and 2. Of course, experience is not always a good guide. For example, an internet app tells me the weight of 1 cubic metre of polystyrene is over 1 tonne. But polystyrene is so light! It does not seem sensible: can this possibly be right? A mathematical literate person will double or even triple check. Answer: yes, it is probably correct, but 1 cubic metre of the polystyrene *foam* commonly used in packaging weighs only 50 kg. Even that is too much for me to lift! A cubic metre is indeed a very large volume.

The two short formative assessment 'smart::test' [3C] items in Figure 3.4.1 involve estimation of length. Real-world estimation involves many skills, each targeted by different items in the length tests and for which teachers receive detailed reports (Stacey et al., 2009; Stacey et al., 2018). The top item in Figure 3.4.1 tests knowledge of how long a millimetre is: students might answer using 'benchmark' knowledge that a finger is about 1 cm wide. The lower item is more complex, involving visualisation, estimates of length in comparison to the ruler shown, and appreciation (from the multiple-choice answers supplied) that rough calculations will give an adequate answer. I counted about 20 horizontal segments, with estimated length of 50 cm (all these lengths equal from the geometry), and four sloping segments of about 70 cm, leading to answer D. Items in a similar style in other smart::tests report on estimation of numerical quantity, area, mass, and volume.

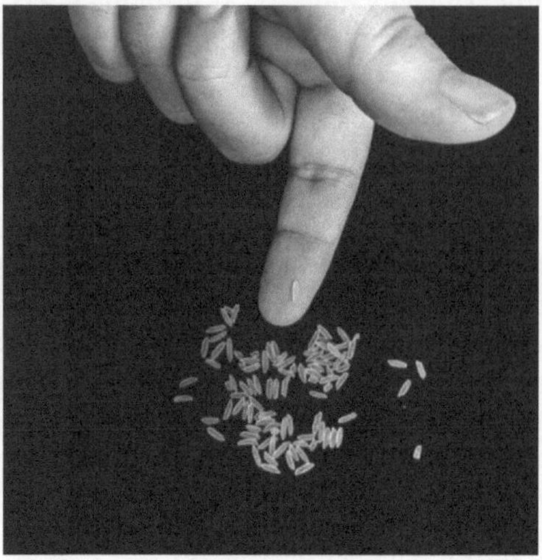

The length of a grain of rice is

A. less than 2 mm
B. between 2 mm and 10 mm (correct)
C. between 10 mm and 20 mm
D. between 20 mm and 30 mm
E. more than 30 mm

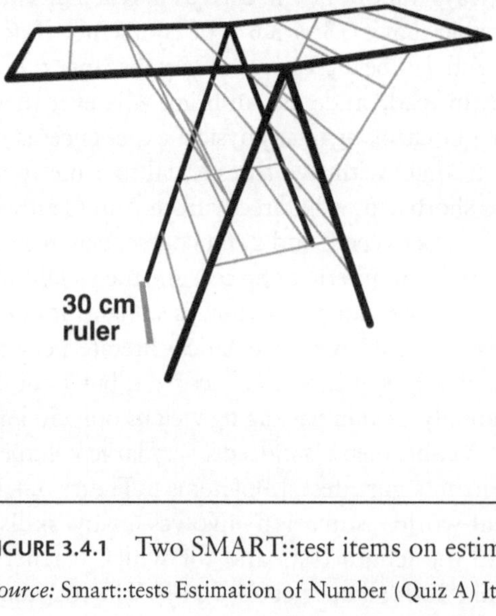

This picture shows a clothes airer made of two different thicknesses of plastic-coated wire. There is a 30 cm ruler on the floor.

The total length of the *thinner* wire is

A. less than 2 metres
B. between 2m and 5 m
C. between 5 m and 10 m
D. between 10 m and 15 m (correct)
E. more than 15 m.

(Modified from original to account for the black & white image)

FIGURE 3.4.1 Two SMART::test items on estimation of length.

Source: Smart::tests Estimation of Number (Quiz A) Items 4151 (top) and 4282 (bottom) [3C].

Classroom discussion of even short items like these can illustrate many points about real-world estimation:

- That the desirable precision in an estimate depends on the real-world purpose of the answer (in these items – detached from a real-world purpose – precision is only indicated by the range of multiple-choice answers provided as in Space Flight task in Figure 2.1.1).

- That speed is important: estimation is used when detailed counts, measurements, and calculations are too slow, not required, or perhaps impossible.
- That the base knowledge for estimation is an appreciation of the size of measurement units and the relative size of numbers, with enough fluency with number facts (including power of ten) and the effect of operations to estimate the results of calculations.
- That other cognitive skills related to the context are needed to contribute to estimation, for example, three-dimensional visualisation for the clothes airer item.
- That a mathematically literate person knows a range of benchmarks for common units of measurement (for example, they might know that the height of a tall man is about 2 metres, that 1 litre of water weighs a kilogram, that a finger is about 1 cm wide, and that a can of drink holds about 400 mL).

Students can learn strategies for estimation, many of which involve proportional reasoning (especially scaling known benchmarks up or down).

Statistics – expanding the range of data displays

Whilst the school curriculum in many countries may still be focused on pictograms, bar charts, and pie charts, new software and new capabilities of digital publishing have prompted the growth of new and more informative data displays. Even if a mathematics curriculum only requires the students know how to *create* traditional data displays, students dealing with data from other areas of interest will need to interpret new graph types.

Figure 3.4.2 presents one example of the new types of data displays that are now being used. It is adapted from the outstanding Gapminder website [3D]. This site presents a huge and constantly updated database of statistics about people and countries around the world. The main messages are that facts (data) are powerful, that things have been getting better over time for most people, and that people around the world who have similar incomes have reasonably similar lifestyles no matter where they live (Rosling et al. 2016). In addition to promoting these powerful messages, the Gapminder website has pioneered and popularised new ways of displaying data. Traditional data displays taught at school (e.g. pie chart or line graph) show the relationship between only two variables, but the outcomes of concern to the Gapminder website are often linked to many variables, necessitating a more complex display. Figure 3.4.2 shows how a 'bubble chart' can combine GDP per capita (horizontal axis), life expectancy (vertical axis), population (size of bubble), and country (label). This figure has been drastically simplified here to meet the limitations of the printed page – the live online version [3D]

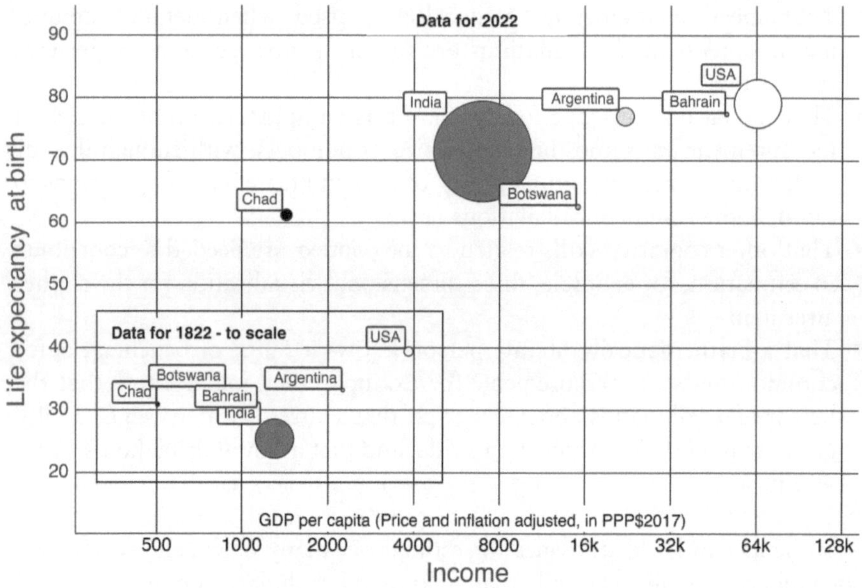

FIGURE 3.4.2 Life expectancy versus per capita income in 2022 compared to 1822.
Source: Free material from Gapminder [3D].

includes far more countries, adds colour to show the world region, and motion to show the passage of time, ultimately combining six variables on one chart, and allows the data for any country to be viewed by pointing to it. This data display powerfully illustrates that life expectancy is strongly linked to average income and that all countries where data is available have improved on both variables over time. Compare the 2022 with the 1822 data of the inset on the figure. Because the 'bubbles' represent country population size, the display also shows that the improvement has been for people, not just for countries. Mathematically literate citizens can easily investigate other relationships of interest by selecting their own variables and data sets to better understand our changing world. Further classroom activities for exploring data are offered in Chapters 4–10 of this book.

3.5 Deep engagement with the context through interdisciplinary work

As we have noted many times above, a distinguishing feature of teaching mathematical literacy must be to help students to intertwine their understanding of the context and the mathematics. Working with teachers of other subjects in a secondary school or using cross-curriculum investigations in a primary school can provide excellent opportunities to deeply intertwine an authentic context with mathematical thinking. Done well, this strengthens students' appreciation of mathematics and its applications ('know about' as well as 'know how'), improves some switched-off students' disposition towards mathematics, and also gives serious attention to the mathematical concepts and skills that are needed in real contexts using the language and practices of the area of application.

To do this well, we need to base lessons and projects on very well-selected topics. Much mathematical work outside of the mathematics classroom is very routine, such as adding numbers, and does not involve significant mathematical thinking. Minimising mathematics removes obstacles for the teacher trying to teach difficult concepts in another field, but it does not assist with developing mathematical literacy. Fortunately, there are many good resources that suggest interdisciplinary lessons with substantial content on both sides (see e.g. Goos et al., 2019); below we describe two such lessons.

Planning for interdisciplinary teaching requires some degree of cross-subject planning. As noted above, it is nearly always important that the mathematical concepts and skills involved are already familiar to the students, because learning about new content will take considerable 'mental space', distracting from focus on the context. Working with teachers of other subject areas has the side benefit of enriching mathematics teachers' own understanding of the different ways in which mathematics is used, and the varying mathematical practices and language that their own students have to learn for different school subjects, disciplines, trades, and jobs.

In summary, even if they are not directly involved in teaching the interdisciplinary lessons, mathematics teachers can encourage and support teachers of other subjects to include significant mathematical content, ensure the relevant content is already well understood, and look for interdisciplinary opportunities that will help their students become more mathematically literate.

Isometric drawing of licorice allsorts

This lesson resource comes from the *Numeracy across the Curriculum* project of the Department of Education in Victoria, Australia [3E]. The website provides a range of sample lessons looking at the mathematical literacy demands of most school subjects for students aged 11–16. Each lesson is prefaced by a short survey of the main mathematical demands of the subject at the relevant stage of the curriculum, which is useful for both mathematics teachers and the subject teacher.

This lesson on isometric drawing is written for teachers of the subject *Visual Communication Design* for students 12–14 years of age. Isometric drawing is a straightforward drawing technique that is useful to introduce students to pictorial drawing prior to teaching more difficult technical drawing techniques. It involves one projection of three-dimensional space onto two-dimensional space – one of the big challenges for artists and map makers throughout the ages. This lesson is not about the rich mathematical history of projections, but directly teaches students the rules for isometric drawing, the skills involved, and how it can have visual impact. Full details of this lesson are available online [3E].

A licorice allsort is a delicious sweet, made of two square layers of licorice sandwiched between three brightly coloured sugar paste layers (see Figure 3.5.1). In the lesson, students look at objects drawn with several different projections and compare the visual impact. Licorice allsorts are good to draw because they have a simple and important three-dimensional shape, they are colourful, and each student can have their own allsort to measure and to view from all angles. Students enlarge the real measurements in the ratio of 5:1, so the drawing fits well on A4 paper. They centre the drawing on the paper using their own measurements and learn how to show the layers of the licorice allsort on the three isometric axes. Students experiment with colours and texture before choosing how to render the layers with coloured markers for their final artwork. They use fine liners to outline the drawing and the border. The finished pictures can be assembled to make an attractive classroom display.

Success in this activity depends on knowledge of multiple mathematical concepts and skills, listed in Figure 3.5.1. In turn, students' understanding of these aspects of mathematics is strengthened by seeing them in action in this important context. Furthermore, a practical demonstration of the usefulness of mathematics in activities students enjoy can help build positive dispositions to learning mathematics.

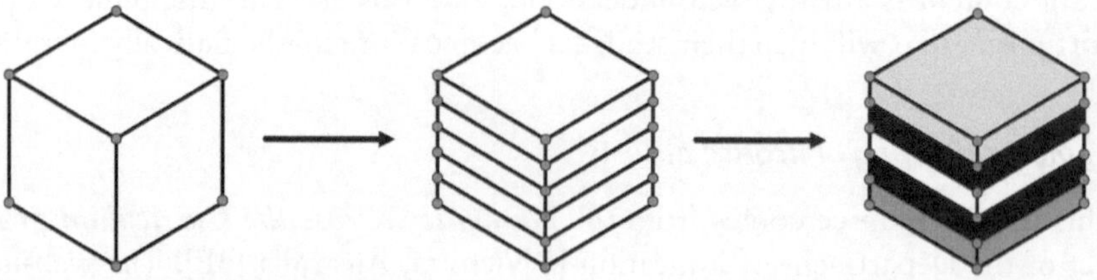

Required mathematical knowledge:

- Knowledge of 2D shapes and 3D objects, including their names
- Identification of parallel and perpendicular lines in two and three dimensions
- Measuring lengths of lines and size of angles (30 degrees and 60 degrees)
- Enlarging shapes in the ratio 5:1
- Using measurements to locate the exact centre of the A4 paper

FIGURE 3.5.1 Stages of an isometric drawing and mathematical knowledge required.

Source: Numeracy across the Curriculum – Victoria Department of Education [3E].

Measuring biodiversity with probability

Biodiversity is important for ecosystem health and productivity. High biodiversity makes ecosystems more resilient to stresses, because having more species means that there is a better chance that some organisms will have traits that are necessary to survive or adapt. To make decisions such as deciding which of several locations is the best place for a marine conservation park or a road, or whether a conservation programme has been effective, ecologists need a practical and reliable measure of the biodiversity of an ecosystem. The lessons highlighted in this section focus on Simpsons' Diversity Index. This is used as a measure of diversity in many biological, economic, geographical, and demographic contexts, and often used in biological subjects in Years 9–12; see, for example, the UK A-level resource website by the Royal Geographic Society [3F]. Because the mathematics behind the index is accessible to students who have studied some probability, it makes an excellent interdisciplinary module, as explained by Duncan et al. (2014). Simpson introduced his index in 1949 and (confusingly) there are several variations in common use, some named after other people. Several online Simpson's Diversity Index calculators are available [3G].

Designing a measure for biodiversity is at heart a challenge to design a mathematical model that captures the key factors, uses obtainable data, and combines the information to produce a number that is meaningful. (This is discussed in Chapter 12.) The module by Duncan et al. (2014) begins by discussing the need to quantify biodiversity, and then illustrates the two key concepts: species richness and species evenness. Species richness is the number of unique species in the area being studied – generally, the greater the number of species, the more biodiverse the area. Species evenness gives a measure of how much variation there is between the number of individuals in each species – generally there is more evenness in more biodiverse areas. Normally, the diversity index is calculated for a group of similar organisms rather than the whole ecosystem, so evenness is calculated from the number of trees of different species (for example), rather than comparing the very disparate numbers of trees at the location with the numbers of ants. To calculate Simpson's index, the species of interest in the area are ascertained and the number of individuals of each species is established (using sampling for large sites). Simpson's Diversity Index is defined as the probability that two organisms selected at random from the area are of different species.

Understanding how the formula is derived is within the mathematical capability of these senior students. Students (probably working in groups) can collect their own data, say to compare two sites, and calculate the indices. Alternatively, they can use published data sets. Students then need to explore the meaning of the index numbers that they have calculated, and how they reflect species richness and evenness. The index is one example of a widespread phenomenon of constructing mathematical functions to act as measures of abstract constructs. Citizens come across these daily. Chapter 12 looks at the mathematical literacy demands this imposes.

This lesson illustrates a number of features about mathematical literacy. Understanding many aspects of the context is essential to understand the goal of the index, why it is needed, and why it works. A teacher of ecology or geography or biology is needed to assist students to understand what variables related to ecosystems might be significant and should be included somehow in the index, how data sets should be assembled, and what data can realistically be collected and how. At first glance, the only mathematics required is to be able to substitute into the formula (or use an online diversity calculator!), but understanding how the formula is derived helps with interpreting the numerical results, as well as demonstrating an important and unusual application of probability theory with a genuine link to students' other studies. Students can explore the results produced by the formula by seeing how extreme situations of richness or evenness influence the index, thereby getting a sense for the strengths and limitations of using such an index (a mathematical model!) for making decisions. Other indices of diversity can be compared. Critical thinking is employed when students provide evidence-based arguments based on the index to support their decisions and to evaluate the arguments of others – and critique the strengths and weaknesses of the index.

3.6 Teaching for critical thinking

Previous sections in this chapter are mostly concerned with what is sometimes called 'functional numeracy': how mathematical concepts and skills can be applied in the solution of everyday or workplace problems. However, meeting the intention that mathematical literacy gives people power in their lives requires more than traditional mathematical skills. People also need the capability to *think critically* about problems and their possible solutions, to carefully evaluate the strength of the evidence and examine alternative possibilities, and to consider the implications of findings from a personal, workplace, or social and community point of view. Mathematical models and the outcomes of using them need to be critiqued considering the variables included and omitted, the relationships proposed between them, the quality of evidence for the assumptions, and the interpretation of results. Critiquing Simpson's Diversity Index as a measure of biodiversity is one example. Students also need the *productive disposition* to do so.

How do we help students develop a critical orientation? This has been investigated by many scholars, including Andersson and Barwell (2021). As for all higher order thinking skills, students acquire experience from active learning pedagogies, where they are challenged to form and explain arguments and justify their conclusions while working on problems with some degree of autonomy. Working with others when analysing information and forming arguments and evaluating conclusions means that more ideas can be considered and ideas can be tested as the work proceeds. Having students present their work to others gives opportunities to practise evaluating arguments and assessing the strength of the evidence presented. For mathematical literacy, this will often include thinking carefully about how well the situation has been modelled with mathematics – how appropriately the assumptions and relationships that have been built into the model capture what is important.

Critical thinking is also likely to be prominent in drawing conclusions from data, especially when they go beyond the actual data that has been presented. Chapter 5 on climate change provides instances of both of these aspects. Because of scientific critique, scientists continually refine their climate models so that they can incorporate more and better input information to improve the models. Critical thinking is also important for governments, business, and citizens to understand how to interpret and act on the outcomes. Class discussion is needed to help students learn from their experiences, with the teacher taking a leading role in highlighting key moments and ensuring that the focus remains on asking and answering sensible and important questions about the context. The later chapters of this book provide many examples where the deeper analysis of context enables informed critical thinking.

Controversial and sensitive issues

A consequence of adopting a critical pedagogy is that teachers will need to manage some potentially controversial or sensitive issues in their classrooms. If students are really to consider the social or personal or environmental consequences of the data and models that they study and are encouraged to see their knowledge as giving them power in their lives, then feelings can run high. Mathematics teachers who are moving into this terrain often find it helpful to talk to teachers of subjects in the humanities or social sciences, who have long faced this kind of challenge.

Thinking critically about mathematical results and their implications will raise issues of values and ethics, and political questions about how society works. If mathematical literacy does indeed support decision-making, teachers can expect that controversial conclusions will sometimes arise about the actions that should follow. Within one class, students who have different experiences and background and different values may well draw different implications from mathematical results. But these conversations are important for helping students become informed and active citizens, surely an important goal of education. The most famous example of a student acting on what they learned at school is Greta Thunberg who began a worldwide movement of 'School Strike for Climate' protests.

Some students may disagree strongly with the course of action others recommend, perhaps heatedly. In dealing with these disagreements, mathematical literacy is best developed by keeping the focus of class discussion on the quality of argument presented: are the premises of the argument sound (including the data), are the steps of the argument logically valid, how does the conclusion of the argument lead to the recommendation for action, and what might the consequences of such action really be.

The Gapminder website [3D], discussed in Section 3.4, is a powerful tool for developing students' 'global awareness' and their knowledge of basic indicators of well-being. In our highly connected and quickly changing world, this is important. But just behind the data lie many controversial issues: the phenomenal success in moving people out of poverty in China balanced against political freedom, lingering effects of colonisation, the economic impact of female illiteracy, the extent of women's rights, and the miserable effects of war are just a few. Data from any one country will raise different, but still controversial, issues. Using real social data in the mathematics classroom and expecting students to think critically about it inevitably means that some of the big debates about society are encountered.

Some issues may be especially sensitive for some classes (if, for example, a classmate has recently died) and are best avoided, whilst being potentially beneficial for most classes. Teachers, knowing their students, make the judgement on this, and perhaps seeking advice from student welfare or health education teachers. Chapter 4 raises various issues of this kind – for example, data shows that the death rate for people aged 15–25 is much higher for males than for females. Discussions on money in Chapters 7 and 9 need to be handled in a way that recognises that different students in a class will have very different wealth, opportunities, and values.

3.7 Knowing about modelling and thinking critically about models

This section draws together some ideas about mathematical models that have underpinned most of the examples in these first three chapters. An important part of the 'know about' component of mathematical literacy is to understand the basic ideas behind mathematical modelling – the process of drawing on an entirely abstract world of mathematical concepts, to make predictions and explain real phenomena in the physical and social worlds. Helping students develop this appreciation, so that they know what can be expected by using a mathematical model, and what cannot, is part of teaching for mathematical literacy. Appreciating what mathematical modelling can do depends on understanding its limitations and its strengths. Only a small proportion of people will ever create any but the simplest of mathematical models, but we are all users of them, so the need is more often to 'know about' than to 'know how'.

A first point to teach is that people encounter many mathematical models even in their everyday lives. Rules for cooking are frequently examples of simple mathematical models. My rule for the absorption method for cooking rice is to add 1.5 cups of water for every cup of rice: easy to remember (a great advantage) and works well, although some people recommend two cups of water instead. Does it make sense to say one of these models for predicting well-cooked rice is right and the other is wrong? Not without looking at the assumptions behind the recipe such as the type of rice and also thinking about the criterion of success – what counts as well-cooked rice for you? Perhaps they are both 'good enough', so a home cook can remember either one.

Students also need to know that many 'apps' that people use every day are built on mathematical models (yes, apps to help with cooking rice are available), some extremely complex. When I use an online navigation app to go to work, the system finds several possible routes, drawing on its extraordinarily detailed map of my town. This map is really a digital model incorporating the geography and the road system. It divides the route into segments, based on road type or speed limit. Then it uses the speed limit and/or real-time data about speed gathered from the phones of other users currently driving to make a recommendation for my route. If I want to go at 6 a.m. on Friday morning instead, the model will replace the real-time data from cars with accumulated average speeds for that time. The model is generally surprisingly accurate, but no one expects it to be perfect. It cannot predict that an accident that might soon block an intersection. When evaluating a model for most situations, we ask whether it is often very helpful, not if it is exactly correct.

Mathematical models are most often invisible to end users. For example, social media platforms sell advertising, and can earn revenue when users click on an advertisement. Platforms employ machine learning to identify detailed profiles (mathematical models) of users who are likely to click on certain types of advertisements. By targeting advertisements to those most likely to click on them (and especially those who will 'like' them), the platform can maximise clicks and hence revenue. For news and information, there is research that indicates that negative reactions (outrage) yield more clicks than positive reactions, encouraging the design of algorithms in the routing models to offer more provocative links. There are claims that this has led to widespread harm by polarising social and political attitudes and spreading pornography. A mathematical literacy critique of a model like this might consider its effectiveness in both identifying the most valuable users (the adequacy of the model) and a social critique of the consequences (intended, unintended, certain, possible, good and bad). The uses and misuses of mathematics in politics, media, and business are discussed in Chapter 7. Critical thinking is part of mathematical literacy.

A mathematically literate person needs to appreciate how much our society depends on mathematical models and to appreciate just a little of the complexity behind some of them. To issue a forecast, a computer program for weather prediction works with an enormously complex model – a set of mathematical relationships between hundreds of variables, with millions of input data from land, sea, and space. Our common experience is that the weather predictions are sometimes 'wrong' (rain might not come when it is predicted) but very 'useful' both for trivial decisions (will I take my coat) and for immensely important decisions (should this town be evacuated tonight because people are in the predicted path of a typhoon). Many people are continuously employed to improve the models to get better forecasts. A good model gives a useful prediction most of the time – perfection is not an option. Another type of mathematical model is to make a measure of an abstract construct for entirely practical purposes, as Simpson's Diversity Index in Section 3.5 does. Gross Domestic Product is another example. Such measures can be used to compare, to track over time, and as a basis for decision-making. They are the topic of Chapter 12.

In summary, we recall the famous saying that "all models are wrong, but some are useful" (Box, 1979); so useful indeed, we now cannot live without them.

3.8 How do we move forward?

In this chapter we have set out implications for teaching that addresses the components of mathematical literacy that we outlined in Chapter 1. The focus on taking the context seriously has been present throughout, with an emphasis on taking a 'critical thinking' approach to data and in using and understanding mathematical modelling and models, whose range, ubiquity, and dependence on technologies is outlined in Section 3.7. This is complemented by the suggestions in Section 3.3 and 3.4 for developing a range of mathematical skills in specific areas like estimation that are especially useful for real-world applications. Sections 3.1 and 3.2 outlined two 'fives': the five key components of the context-focused mathematical framework that learning activities must address to develop mathematically literate thinkers within the broader pedagogical context of TRU's Teaching for Robust Understanding, with its five dimensions.

For many teachers of mathematics, this agenda represents a qualitative enlargement of the range of learning goals to which they are accustomed. They can form part of ongoing professional development through local discussion, formal professional learning sessions, or community models such as lesson study. There is online support for such professional development, such as the Bowland Maths professional development modules [3H]. These offer a sequence of structured activities for a group of teachers to work together on specific pedagogical challenges. The design is a three session 'sandwich model' of classroom-activity-based professional development, designed to maximise 'carry over' from the sessions to the classroom – worlds that can sometimes seem far apart. Teachers first work together to plan a lesson supported by supplied materials, then teach and observe the outcomes at school, and then come together again to reflect on the activity and what they have learned. The module handbooks describe the activity sequence in detail and address topics such as tackling unstructured problems, assessing modelling processes, and managing collaborative work.

Acknowledgements

Figure 3.4.2 is based on free material from www.gapminder.org available under the Creative Commons Attribution 4.0 International licence. Figure 3.5.1 is adapted from material from *Numeracy across the Curriculum* © State of Victoria (Department of Education and Training), also available under the Creative Commons Attribution 4.0 International licence. The image from Bowland Maths (Figure 3.2.2) appears courtesy of the Bowland Maths maintainers. Material from smart::tests in Figure 3.4.1 appears with permission from Kaye Stacey.

References

Andersson, A., & Barwell, R. (2021). *Applying critical mathematics education.* Brill.

Box, G. E. P. (1979). Robustness in the strategy of scientific model building. In R. L. Launer & G. N. Wilkinson (Eds.), *Robustness in statistics* (pp. 201–236), Academic Press. https://doi.org/10.1016/B978-0-12-438150-6.50018-2.

Brousseau, G. (1997). *Theory of didactical situations in mathematics: didactique des mathematiques, 1970–1990.* Kluwer.

Brown, J. P. (2019). Real-world task context: Meanings and roles. In G. Stillman & J. Brown (Eds.), *Lines of inquiry in mathematical modelling research in education.* ICME-13 Monographs. Springer. https://doi.org/10.1007/978-3-030-14931-4_4

Burkhardt, H. (2021). Modelling in school mathematics: Past achievements – current challenges. In G. A. Stillman, W. Blum, & G. Kaiser (Eds.), *Mathematical modelling and applications, ICTMA 19* (pp. 529–539). Springer.

Duncan, S. I., Lenhart, S., & Sturner, K. K. (2014). Measuring biodiversity with probability. *The Mathematics Teacher*, *107*(7), 547–552. https://doi.org/10.5951/mathteacher.107.7.0547

Goos, M., Geiger, V., Dole, S., Forgasz, H., & Bennison, A. (2019). *Numeracy across the curriculum: Research-based strategies for enhancing teaching and learning* (1st ed.). Allen & Unwin.

Madison, B. L., & Steen, L. A. (Eds.), (2008). *Calculation vs. context: Quantitative literacy and its implications for teacher education.* Mathematical Association of America. https://www.maa.org/sites/default/files/pdf/QL/cvc/CalcVsContext.pdf

Phillips, R., Burkhardt, H., Fraser, R., Coupland, J., Pimm, D., & Ridgway, J. (1988). Learning activities and classroom roles with and without the microcomputer. *Journal of Mathematical Behavior*, 6, 305–338.

Rosling, H., Rosling, O., & Rönnlund, A. R (2016). Factfulness: Ten reasons we're wrong about the world – and why things are better than you think. Sceptre.

Shaughnessy, J. M. (2007). Research on statistical learning and reasoning. In F. K. Lester (Ed.), *Second handbook of research on mathematics teaching and learning* (pp. 957–1009). Information Age Publishing.

Stacey, K., Price, B., Steinle, V., Chick, H., & Gzodenko, E. (2009). *SMART assessment for learning* (paper presentation). International Society for Design and Development in Education Conference 2009, Cairns, Australia. Retrieved from https://www.isdde.org/conferences/conference-cairns-2009/working-groups-and-themes/curriculum-documents-assessment/

Stacey, K., Steinle, V., Price, B., & Gvozdenko, E. (2018) Specific mathematics assessments that reveal thinking: An online tool to build teachers' diagnostic competence and support teaching. In T. Leuders, K. Philipp, & J. Leuders (Eds.), *Mathematics teacher education: Diagnostic competence of mathematics teachers* (pp. 241–261). Springer.

Steen, L. A. (Ed.). (2002). *Mathematics and democracy: The case for quantitative literacy*. National Council on Education and the Disciplines.

Stillman, G. (1998). The emperor's new clothes? Teaching and assessment of mathematical applications at the senior level. In P. Galbraith, W. Blum, G. Booker, & D. Huntley (Eds.), *Mathematical modelling: Teaching and assessment in a technology-rich world* (pp. 243–253). Horwood.

United Nations Sustainable Development Goals for 2030 (n.d.) *Target 4.6*. Retrieved from: https://sdgs.un.org/goals/goal4

Links to useful material

To visit any of these links, scan this QR code or visit ltml.mathlit.org – append the link code to go directly to the entry – e.g. ltml.mathlit.org/3A

The original source links are given below for attribution purposes.

[3A] *The TRU Framework* – Teaching for Robust Understanding Project
https://truframework.org/

[3B] *Reducing Road Accidents* – Bowland Maths
https://www.bowlandmaths.org.uk/projects/reducing_road_accidents.html

[3C] *Estimation of Length and Mass Items* – SMART::tests
https://www.smartvic.com/

[3D] *Understand a Changing World* – Gapminder
https://www.gapminder.org/

[3E] Isometric Drawing – *Numeracy across the Curriculum*
https://www.education.vic.gov.au/school/teachers/teachingresources/discipline/maths/Pages/numeracy-for-all-learners.aspx#link27

[3F] *A Guide to Simpson's Diversity Index* – Royal Geographical Society https://www.rgs.org/schools/resources-for-schools/a-student-guide-to-the-a-level-independent-investigation-non-examined-assessment-nea

[3G] *Simpson's Diversity Index Calculator* – Statology https://www.statology.org/simpsons-diversity-index-calculator/

[3H] *Professional Development Modules* – Bowland Maths https://www.bowlandmaths.org.uk/pd/

4

HOW RISKY IS LIFE?

Risk is an area of life that concerns many people – and rightly so. Numbers are obviously involved, so it is an important aspect of mathematical literacy. But risk is also an area where many people's awareness and concern have little connection to the level of risk actually involved in a situation or an activity.

This sometimes has life-changing consequences. Parents are reluctant to let their children 'play out' which undermines their physical and mental health and development. Inadequate physical activity reduces general fitness and increases the risk of obesity and diabetes – screen games are no substitute. People have a fear of flying or other phobias, where the effects on their career and lifestyle may be significant. Yet the data shows that the likelihood of harm in such cases is extremely small. These are just two examples where increasing mathematical and, in particular, data literacy can improve quality of life.

These are emotional subjects that make it hard for deliberative 'slow thinking' to cut through instinctive reactions to news reports of tragic events, and thus to improve decisions. The media know that people like stories – but stories are not data from which you can make reliable judgements. Any discussion of risk must take this into account.

The theme of this chapter is that people can make better-informed decisions about what to do about potentially risky situations if they take the trouble to make informed estimates of the likelihood of the risk. With a bit of searching, adequate data is often available on the web. Learning to do this kind of deliberative thinking, and being confident in doing it (the productive disposition component of the context-focused mathematics framework of Chapter 1) is an aspect of mathematical literacy with a big potential payoff in quality of life – one that many find interesting, and motivating.

DOI: 10.4324/9781003303503-5

In Section 4.1 we set out a general framework for thinking about risk, moving on in Section 4.2 to describe an approach to developing understanding of risk in the classroom, based on well-engineered materials that are freely available online. Section 4.3 looks at the challenges of finding reliable data and probing it for other undeclared influences. The COVID-19 pandemic made people more aware of risk – the chance of getting infected, and of the consequences. Section 4.4 looks at this complex example in some detail, pointing out general features and lessons learned that may help preparation for the next epidemic which, with unknown effects and mechanisms, is expected in a decade or so. Section 4.5 looks at the issues in medical testing – an example of conditional probability directly relevant to the mathematical curriculum where there are surprising results. Again, effective lesson materials are available free online. The chapter includes a mix of 'know how' and 'know about' the analysis of risk, using a critical thinking approach. It also provides many examples of using both the theory-driven and data-driven approaches to understand a situation. Risk is a topic, that interests almost everyone, helping to build a productive disposition to investigation in general.

4.1 Dimensions of risk

There are two dimensions of risk that need to be distinguished. Any potential event has both a likelihood and a level of hazard:

- *Likelihood:* How likely is this to happen?
- *Hazard:* How serious is it, if it does happen?

These two aspects both need to be taken into account, and balanced, in making decisions. In this chapter we use 'likelihood' as the general term, reserving 'probability' as a measure of likelihood in its formal mathematical sense – a number in the range 0 (never happens) to 1 (for sure).

Some events are high on likelihood but low on hazard. I often forget where I left my keys –but I only waste a few minutes searching for them. I'm likely to get a cold this winter, but it usually only means a few unpleasant days. So I won't change my lifestyle to reduce the likelihood.

Other events are very unlikely but disastrous if they do occur. The chance of a large meteorite hitting the earth is extremely small, but the one that plunged into the Gulf of Mexico 66 million years ago sent so much stuff into the atmosphere that the whole ecology was changed for long enough to kill off the non-flying dinosaurs (others became birds) and create ecological 'space' for mammals to increase and diversify – which might be seen as progress. Scientists now track objects in space with a view to diverting any that pose a risk like that. On hazard at a personal level, each of us has in mind possible life-changing events that we regard as disastrous – someone close to us dying is an obvious example; we take what steps we can to lower the likelihood of that happening.

Most potential events are in between these extremes – with fairly low likelihood and consequences that are unpleasant but not life-changing. Risking breaking your ankle on a school outdoor adventure trip, or a parachute jump, or riding your bicycle to school every day to improve your fitness present risks; conversely, not exercising regularly is also a major risk to health – as is spending every evening playing video games. These choices all have risks to health and well-being and there is data that enables you to estimate their likelihood, at least for the average person.

These examples illustrate another distinction that is important in thinking about risk: between 'one-off events' like the parachute jump and ongoing activities like riding a bicycle to school.

Because quantifying hazard involves subjective judgements, people differ in how serious they regard the hazard to be. In much of this chapter. we shall 'fix' the hazard variable by concentrating on the likelihood of various *fatal* outcomes. Death also provides a background frisson in classroom discussions that, though it needs sensitive handling, students seem to enjoy.

Measures of risk to life

R.A Howard (1980) introduced a useful unit for comparing risks to life, the *micromort*: a one in a million chance of death – or probability $0.000001 = 10^{-6}$. It is a useful unit because long decimals are hard to understand.

ACTIVITY

Discussing in small groups, conjecture how many micromorts (chances in a million) is the average likelihood of dying in: a parachute jump; giving birth; a car accident during one year.

Table 4.1.1 gives some examples for individual events, and for ongoing activities like riding or driving in wealthy countries.

Base risk

Another important number in thinking about risk is the total probability that an average person in a specified group will die of any cause in the coming year. We call this 'base risk'. The most important variable affecting base risk is age – though many other variables, including health, gender,

TABLE 4.1.1 Likelihood of dying in various ways in wealthy countries

Hazard	*Chance of death*
Single event	
A skydive/parachute jump	8 micromorts
Climbing to the top of Mt Everest	40,000 micromorts
Giving birth	150 micromorts
Baby's first day of life	430 micromorts
For ongoing activities	
Being murdered	10 (UK) to 50 (US) micromorts per year
Car accident	25 micromorts per 10,000 km
Travelling by train	1 micromorts per 10,000 km
Motorbike accident	1000 micromorts per 10,000 km
All unnatural causes	230 (UK) to 460 (US) micromorts per year

Source: Data from Wikipedia and other sources.

socio-economic status, occupation, and country, will influence it, sometimes substantially. Base risk as a function of age is a component of the 'life tables' published by national statistics authorities. (You might think 'death tables' is a better name!) Table 4.1.2 is a recent example for England and Wales, also showing the effect of gender. The last column gives the corresponding expected age at death – numbers that are perhaps surprising at first but obvious on reflection.

TABLE 4.1.2 Probability of death in coming year, and expected age at death

Current age	*Probability (micromorts)*		*Expected age at death*	
	Males	*Females*	*Males*	*Females*
0	4,220	3,500	79	83
1	229	214	79	83
5	74	74	79	83
10	78	66	79	83
20	525	187	80	83
30	771	387	80	83
40	1610	924	80	84
50	3580	2,230	81	84
60	7910	5250	83	85
70	19 200	12 900	85	87
80	56 100	39 600	88	90
90	163 000	136 000	94	95

Source: Calculated from UK Office of National Statistics tables England and Wales 2018–2020 [4A].

It may seem strange that the chance is not higher at the older ages – only about 15% (150 000 micromorts) chance of dying within a year at age 90. But we can estimate, say, the 5-year survival probability at age 90. The probability of surviving 5 successive years $(1 - 0.15)^5 = 0.44$, which makes it plausible that 95 is the expected age of death.

One of the interesting features of the COVID-19 pandemic was how closely the chance of death from a COVID infection paralleled the base risk as a function of age – increasing by a factor of around 2.5 for every 10 years from age 30.

In considering how seriously to take a risk, your base risk is a useful comparator. Not that we are recommending parachute jumping for 80-year-olds – but, if you fancy it, sky-diving is a relatively minor risk.

The base risk of dying and the danger of various hazards provide the background to answering one of the questions with which we began this book:

> What is the chance that a teenager (like my child) will die in a shooting at school? How does it compare to other hazards in their life?

In 2022, 32 students were killed in 51 school shootings in the United States. There are 115 000 US schools that educate 50 million students across the age range 4–18. So the risk of dying in a school shooting in 2022 was less than 1 micromort. School shooting deaths are a very small part of

- the 30 000 shooting deaths each year;
- murders (the average murder risk in the United States is 50 micromorts (see Table 4.1.1); or
- 'natural' deaths of school-age children as suggested by the UK figures in Table 4.1.2. (One should check the US numbers, of course.)

The chance of having a shooting in 'your school' during the year 2022, a traumatic event for the whole school and its community, was roughly 1 in 2000, averaged across the United States. Over the total period of schooling of 10+ years, it is around 1 in 200. Whilst these risks are far too small to outweigh the benefits of school attendance for an individual, they are sufficiently large to identify this as a problem of national significance.

Measures of likelihood

We conclude this agenda-setting section with a foretaste of the teaching challenges of reconciling the formal mathematics of probability with comprehensibility in this area. The fundamental concept, probability, measured with a number in the range 0–1, proves very difficult for people to grasp and, particularly, to use. (No surprise for teachers there.) This is exacerbated by the very small numbers often involved in risk estimates – 8 micromorts is a probability of 0.000008. This difficulty is not confined to students in school; there is a large body of evidence showing this, across a number of professions, including doctors (Hoffrage et al., 2000), particularly when comparative probability is involved. Across media it regularly leads to eye-catching headlines like "Research shows that ... doubles the risk of developing cancer of the". But if the cancer is rare, then double the risk is also very low.

It is now well established that communicating about risk is most effective when done in terms of numbers in a specific size of population; these are sometimes known as 'natural' or 'expected' frequencies. Large whole numbers, though challenging to visualise, are more accessible than lengthy decimals. For this example, if 1 person in a 1000 develops the cancer, doubling the risk is not too serious; if the normal level is 100 in 1000, that is another matter. Micromorts are, of course, the expected number of people dying from a population of 1 million.

All this evidence strongly suggests that expressing probabilities as numbers in a specified population is a good approach to developing mathematical and statistical literacy. The classroom trials in the development of the online teaching unit 'How risky is life?', described in the next section, reflected these results.

Appropriate precision

Risk is also a good area to bring out the important concept of *appropriate accuracy*: that a calculated number cannot be more accurate than the assumptions made and the numbers used in the calculation. For risk this often means the numbers are only 'ballpark estimates' and more so because the input data may be based on populations that are significantly different from those where the conclusions will be used, which might often be just for an individual. Mathematical literacy usually only requires rough numbers to guide decisions – hence the numbers in Table 4.1.2 are rounded.

Appropriate degree of precision is important in all applications of mathematics (see Chapter 3), but it rarely gets much attention in traditional school mathematics. It is not unusual to see nine digits, copied off a calculator as an answer to a real-world problem, go unchallenged by the teacher. Lessons focused on mathematical literacy are a good place to address this issue, because knowledge of the context can provide clues to the accuracy of the data.

4.2 Exploring risk in the classroom

In this section we describe an approach to risk that tackles many of the issues set out above in a five-lesson unit aimed primarily at 14–17-year-old students. Designed and developed by the Shell Centre and reSolve teams, two versions of the teaching materials are available free online [4B]; each gives detailed teaching suggestions and supportive tools. The first, from *Bowland Maths*, focuses on risk, while the second, from *reSolve: Mathematics by Inquiry* (Australian Academy of Science), also addresses the modelling aspects more explicitly. Feedback from classroom trials indicated that the learning and enjoyment goals both were achieved by these sequences of lesson activities, which we outline in this section. (Readers who are interested in design and development may find it interesting to compare these two versions, developed by overlapping teams for two different communities of teachers and students.)

Both "How risky is life?" units focus on a specific hazard (death) and its likelihood, starting with deaths from unnatural causes – often a major source of anxiety. Here we give an overview of the structure of these modules, reviewing the essential insights that each lesson addresses.

Impression versus reality

In preparation for the unit, the teacher encourages students to bring various media reports of deaths into the first lesson, *Exploring our perceptions of risk*. Groups of students rank their impressions of the probabilities of different unnatural causes of death. The different groups' views are then discussed across the class and a consensus arrived at.

The second lesson, *Sudden death – what does the data say?* presents the real data on the number of deaths per year for ten different causes, aiming to bring out the mismatch between 'common sense' and reality in risk. Three simple visual tools (designed by Malcolm Swan) are used to establish a (logarithmic) scale of likelihood based on whole numbers, expressed as "1 in 10,000" etc., aiming to emphasise how small these risks are.

The first tool is a printed sheet with strips for each decade of likelihood, described as "1 in ..." and in words, where the students calculate the numbers of deaths in the whole population for each cause (using ballpark figures of 25 million for Australia and 50 million for England). The second tool is a poster that each group prepares using the real data and the likelihood strips, sequenced in descending order of risk. Figure 4.2.1 shows a successfully completed UK poster. Note that the higher risk strips are not required. Again, the numbers will be different for different periods – for example, the terrorism number mainly reflects one incident.

The third device, also looking forward to the next lesson, helps students to visualise the *absolute size* of each risk of death. It is simply a sheet of paper with 100 squares in a 10×10 array, and then each such 'large square' is further divided into 100 'small squares'. Students first calculate how many people in the population each small square represents (5000 in the England case), assuming the page represents the whole population. Then each student colours in an area corresponding to each unnatural cause of death. In total less than two small squares will be coloured, representing less than 200 micromorts per year – too tiny to reproduce legibly here. The diagram can be found on the website [4C].

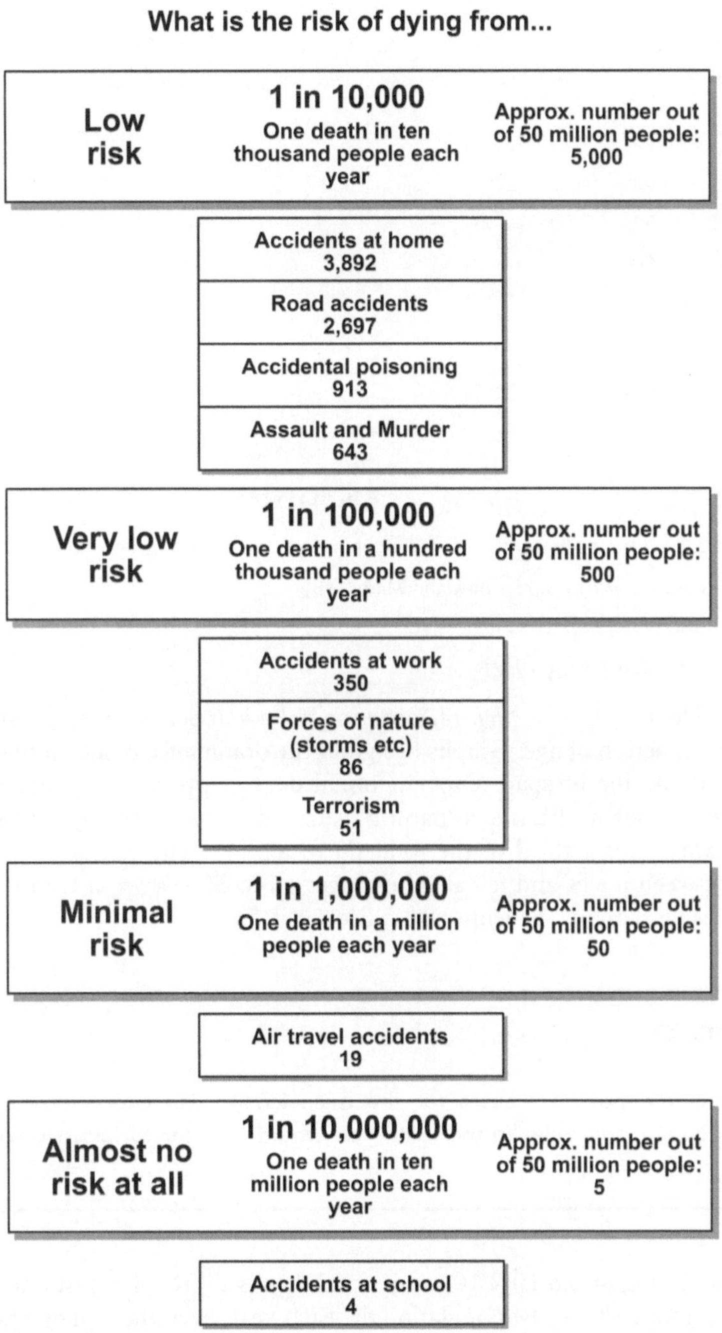

FIGURE 4.2.1 Likelihood of deaths from various unnatural causes (UK).

Source: From *How risky is life?* – Bowland Maths [4B].

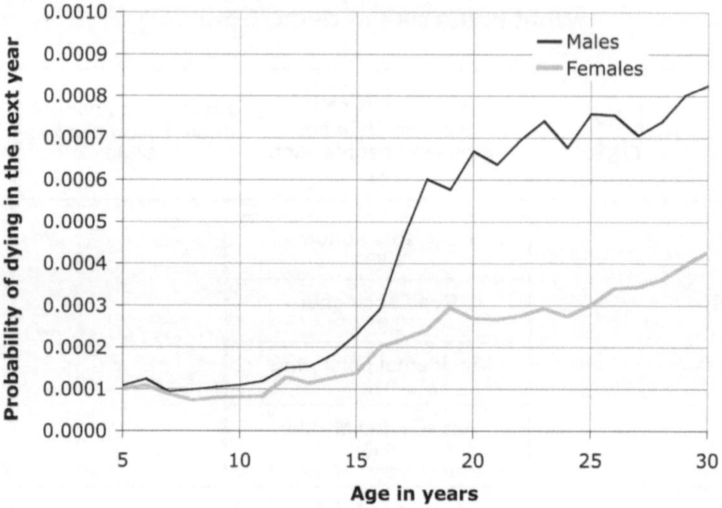

FIGURE 4.2.2 The male–female differences in base risk among younger adults.
Source: From *How risky is life?* – Bowland Maths [4B].

Risk of death from all causes

The third lesson, *The big picture*, moves on to look at deaths from all causes – now as a function of age. Graphs bring out the dominance of age in the scale of risk across the lifespan. Other national data (supplied in the materials) shows that poor health (i.e. a 'natural' cause) dominates, though unnatural causes are about a third of the total up to age 40. The substantial differences between males and females from ages 15 to 30 shown in Figure 4.2.2 promotes discussion, as mentioned in Section 1.5.

ACTIVITY

Discuss the possible causes for the male–female differences shown in the figure – and what further data you would look for to improve your understanding.

Again, filling in the 10 × 10 array, total deaths fill roughly just one large square – about 1% of the population die each year, overwhelmingly at older ages as expected. So life is risky but death is very unlikely – until old age where, if you get there, as we saw in Figure 4.1.2 you're still likely to survive a few more years.

The lesson goes on to present more detailed data on the way different causes vary with age, involving translation between different representations. It then goes on to probe understanding with a collection of statements to be classified as 'True, false, or impossible to tell?' from the data provided.

Deeper issues of data literacy

The unit then goes on to explore four important data literacy concepts in a little more depth. Centrally important is not taking data at face value but critically probing the assumptions behind it. We then explore the mathematics of random variation, going on to discuss how far one can draw inferences for individuals from population data, and the concept of 'best estimate'.

Probing assumptions behind data

Going further, Lessons 4 and 5 of the reSolve unit explore these issues and relate them to the process of data-driven modelling. Lesson 4 aims to help students avoid just taking data at face value (even data from very reputable sources), carefully considering what conclusions can really be drawn from it, thinking critically about the situation and asking questions.

One activity is to decide if men are more dangerous drivers than women. The data in Figure 4.2.3 is from the national statistics on cause of death in Australia for 2015. This data clearly shows that at every age up to 85,

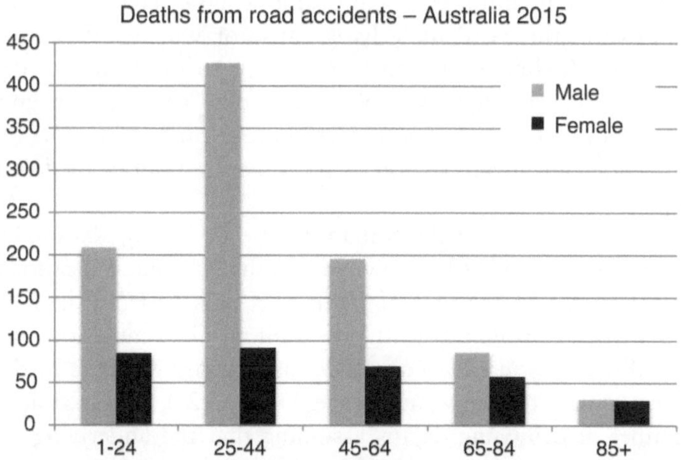

FIGURE 4.2.3 Number of deaths from road accidents in Australia in 2015 by age and gender.

Source: Graph adapted from the reSolve *How risky is life* module [4B].

more men than women die in road accidents. Maybe men *are* more dangerous drivers than women, but they become more careful in very old age.

Students, working in groups to critically analyse the situation, will quickly come up with reasons why the conclusion of dangerous driving cannot be deduced from this data. The data includes road deaths for drivers, passengers, and pedestrians. It does not account for how far or for how long men on average were driving or were passengers in 2015 compared to women, nor the possibility that men may more often have to drive in more dangerous circumstances. Equally, change could be explained by there being far more women than men over 85s, and hence more women over 85 driving. One needs rates! We do not even know from this actual data that there was a similar pattern in other years (although there is). Further information on many additional variables has to be considered before we can decide whether men are more dangerous drivers than women. Of course, in real life, it is very likely that there will not be sufficient data available to make a decision on any one specific question; and the best that can be done is to make reasonable inferences from all available data, acknowledging limitations. This activity is the heart of critical thinking, taking seriously the context and all its variables.

Random variation

reSolve Lesson 5 is about making predictions from data. It emphasises that data (such as the number of road accident deaths) varies from year to year because of random fluctuations and also because of 'real' changes that are likely to persist. Simulation can help decide whether changes in data arise from a change in the 'real' underlying cause or whether they are due to random factors. A simulation app (a precursor to the study of confidence intervals providing concrete examples) is used to show the magnitude of variation in data that can be expected from these 'random' factors that have no known significant meaning. The app can be used online or downloaded from [4E].

For example, in 2017, the Australian Royal Life Saving National Drowning Report [4D] reported 291 drowning deaths in aquatic locations across Australia. That was an increase of 10 deaths (or 4%) from the rolling 10-year average of 281 drowning deaths per annum. The simulation app can be used to see if it is reasonable to conclude that the risk of death by drowning really increased over the ten-year average. Figure 4.2.4 shows 20 simulations of the number of drowning deaths assuming that the underlying probability of dying this way is 281 in 25 000 000 (the approximate population of Australia). In that case, 11 of the 20 simulations gave more than 290 deaths per year. When the simulation was repeated 5 000 times, it showed that over 290 deaths would happen in about 40% of years from these random causes,

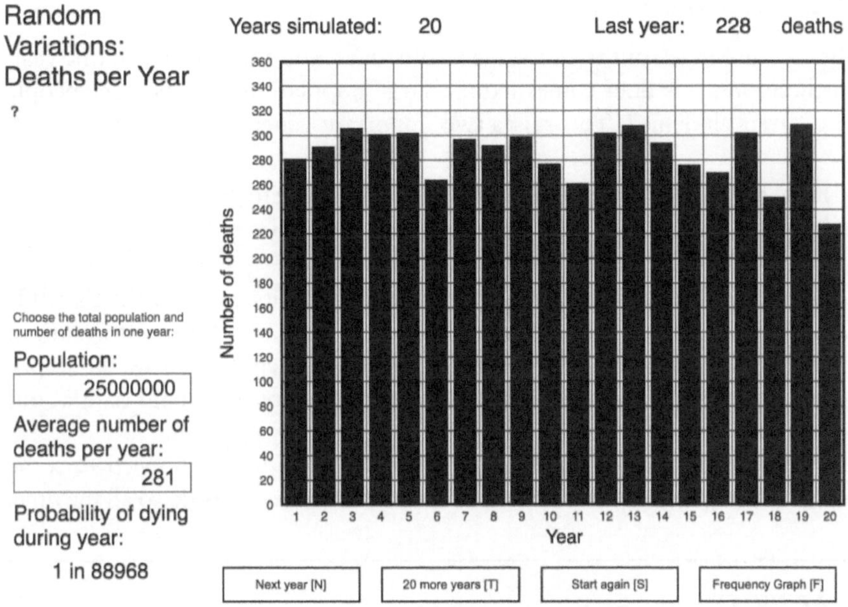

FIGURE 4.2.4 Simulated number of drowning deaths over 20 years assuming a fixed risk.

Source: App by Shell Centre for Mathematical Education – used in Bowland Maths & reSolve [4E].

without an actual change in the likelihood of drowning. The high number of deaths (291) is disappointing when the desire is to reduce drowning, but getting such a high number is easily able to happen by chance. There is no evidence from this data that the underlying chance of a person drowning in Australia had really increased in 2017. Understanding variation in data and being able to put some measure on it is essential for interpreting data, and hence for mathematical literacy. And a sad postscript: data since 2017 shows steady increases.

From looking at multiple (simulated) examples, students are encouraged to observe that $\sqrt{n}$ is usually a useful rough estimate of the random variation to be expected in a situation where there is a rare event occurring about n times in a given time period. (Here about 280 deaths per year; another example is number of traffic accidents per year.) This rule (which comes from the Poisson distribution) is an example of useful 'know about' mathematics since most people never study the Poisson distribution. Compared with other subjects, mathematics is reluctant to include 'know about' results, insightful though they can be. For the 281 drownings, $\sqrt{n}$ is about 17, so 10 is well within expected random fluctuations.

What can I infer about me?

The data use in lessons to this stage is 'old' data, and some from just one year. Furthermore, this data comes from a large population that includes people who "aren't like me". This raises two important general questions:

- How far does this data still apply *now*, several years later?
- How far does this data apply to *me*, an individual?

The aim of the guided exploration is that students come to see that this data provides a *best estimate* from the data we have. More information, involving more variables than just age and gender, could narrow the population sampled and allow a better estimate, but there will be no data about populations of people exactly like you.

The other big question is what would you do with the knowledge. Some people won't take any risk that they know about, while others do so all the time.

> **ACTIVITY**
>
> The PISA task **EARTHQUAKE** (M509 from the 2006 released items [4M]) assesses students' basic ability to interpret probability statements (as do many other PISA items).

4.3 Finding reliable data – critical thinking for data literacy

In this section we look more broadly at the challenges of data literacy – in particular, how students can learn to find and sort through the data they need from the flood of seemingly relevant data that is now available online. This critical thinking is at the heart of data literacy. We explore it in the context of risk, but the approach is relevant wherever data is needed, whether in data-driven or theory-driven modelling approaches.

We start from the position that mathematical literate people have busy lives. They do not want to devote large amounts of time to analysing things that are not critical to their current decision-making. Time-effectiveness in acquiring information is central. Where does one begin?

- **How do you find information on an issue that interests you?**

  This is relatively straightforward. You start by putting the area of your interest into a search engine, trying various wordings. You will be faced with far too many links, often literally millions, starting with paid-for 'Ad' links (normally best avoided). This forces you to think about what the important variables are. As you begin to formulate a model of the phenomenon, you ask: 'is this data a good measure for what we are interested in, in this particular context' – sometimes called 'internal validity'.

- **How do you probe how trustworthy the data is?**

  First, one is looking for reliable sources that have no inherent biases – though no writing is value-free, some is more obviously 'selling' a point of view. Wikipedia is usually a good place to start; it is monitored by a community that feels strongly about objectivity, with clearly specified sources. If in doubt, 'digging deeper' on Wikipedia will also reveal community discussion on the accuracy of the article.

  Recently, 'artificial intelligence' systems, such as ChatGPT, have become available to provide extended responses to questions posed in plain language. At the time of writing, however, these sometimes produce plausible sounding but completely fictitious answers. In one early case, a lawyer used such a system to produce an impressive-looking legal argument, but it cited past cases that simply didn't exist!

  However reliable the source, two questions must always be asked. What other influences might affect this data? What other variables should be considered? The dangerous driving discussion above exemplifies these issues.

- **What kinds of data are relatively straightforward to interpret?**

  This is at the heart of data literacy, and its relation to modelling. Ideally one looks for data from a reliable source where only one variable is changing throughout the data set. This data can be compared to a model of the underlying mechanisms, if you have one. Most data is, in contrast, 'messy' with several variables changing across the data set. Modelling such situations is thus more complex, usually involving a more heuristic approach, with parameters determined from experiment.

- **How far are the results generalisable to the situation I am interested in?**

  This question arises whenever, we want to make inferences from data that comes from populations or situations that are different from the one we are interested in. Is this generalisation warranted? Called 'external validity', this can only be assessed by thinking critically through all the other variables that might be significant in the phenomenon you are studying and informally modelling the likely influence of each. Such critical thinking is at the heart of mathematical literacy. In Section 4.4 we will exemplify these general issues in the context of epidemics.

- **Where to find data**

  National statistics agencies (such as the Singapore Department of Statistics) and organisations such as the World Bank, UNESCO, Gapminder and even the CIA publish large data sets online and are usually highly reliable. These can be used to explore a wide range of topical social and economic issues, either with the online visualisation tools supplied by the organisation or by downloading the raw data.

4.4 Epidemics – particularly COVID-19

The COVID-19 pandemic brought discussions of risk from academic journals aimed at experts onto the front pages of newspapers. Since serious epidemics occur every decade or so, understanding the dynamics and the meaning of the multiple kinds of data produced, and their implications for the decisions that you as an individual make, are an important aspect of mathematical literacy. We shall use this as a case study to illuminate some general issues of data analysis and modelling and the challenges it presents in complex situations that are changing rapidly. We look first at the fundamental dynamics of epidemics, moving on to some intriguing aspects of COVID-19.

The epidemic curve

We begin with a standard simplified model of epidemics, which students can explore using the spreadsheet [4F].

The basic effects are two: exponentially growing spread and increasing immunity. Consider a closed population of people who interact with each other. It is a reasonable model assumption, supported by evidence, that when few people are infected, each infected person infects, on average, a given number, say R_0, other people. The infection will then grow exponentially: if $R_0 = 3$, the number infected in each step of the developing infection is 1, 3, 9, 27,.... But it then follows that there is a growing number of people who *have been* infected $1+3+9+27+\cdots$. Assuming that a person cannot be infected twice (increasingly proving not to be the case with COVID-19 and its variants), an increasing proportion of the people an infected person interacts with will have become immune. Consequently R, the average number of people infected by an infected person, decreases from its initial value R_0. When R drops below 1, there is an exponential *decrease* in the number of new infections and the epidemic dies out. The population is said to have acquired *herd immunity*. This behaviour can be explored by varying R_0 in the spreadsheet. This behaviour is shown in Figure 4.4.1.

> **ACTIVITY**
>
> Use the spreadsheet by varying the parameters to explore the level of infection. Watch out for values of R_0 that produce chaotic behaviour caused by step-by-step 'discrete mathematics' effects.

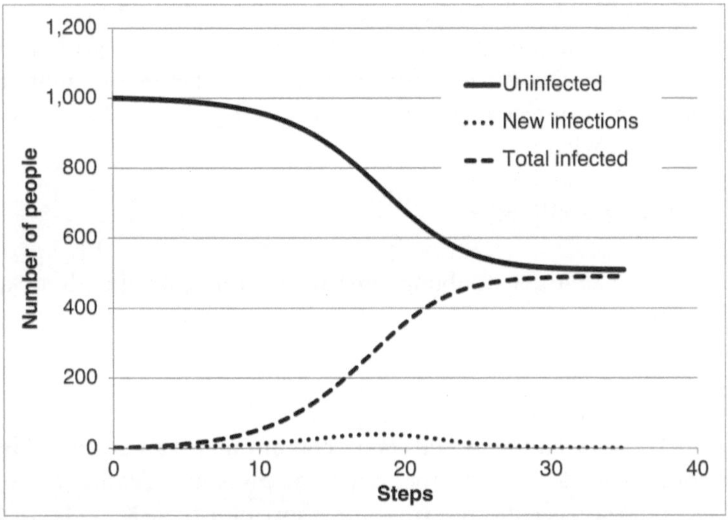

FIGURE 4.4.1 Infection numbers in the classical model of an epidemic.

Source: Population of 1,000 people with $R_0 = 1.3$ – see the spreadsheet at [4F].

R_0 depends on many factors: the inherent transmissibility of the disease, the time a person remains infectious, the average frequency of contact of individuals, the time spent together, and other factors – all integrated across the whole population at risk. All this makes calculating R from first principles complex, though simulation modelling gave useful results throughout COVID. In practice, R_0 was determined empirically from the speed of the initial exponential growth of infections in the epidemic. Beyond its natural contagiousness, R_0 for COVID was increased by two features of the disease: people were infectious for several days before they showed symptoms, and some infectious people never developed symptoms at all.

This is an example of two complementary approaches to modelling described in Chapter 1: theory-driven and data-driven. Theory-driven modelling identifies the relationships between variables, but data-driven modelling works, for example, from graphs of empirical data – a mathematical representation of a phenomenon with no underlying assumptions beyond the choice of variables deemed to be of interest. For most situations a mixture of approaches is appropriate. For epidemics we have a theory-driven understanding of the initial exponential growth and the transition to herd immunity but the values of the key parameters, R_0, and the timescale of each 'step' are to be found from observation in the initial stages of the epidemic.

How are the 'steps' in the transmission of the disease related to time? Often through a parameter that quantifies the rate at which new cases are arising, usually labelled r. This parameter r depends on R and the timescale over which each step of infection occurs. A more accessible concept is the doubling time of the epidemic t_d. With an exponential growth model (number infected at time $t = ke^{rt}$) for the early stages before immunity develops, the number of infected people at time $t = 0$ is $ke^{r\times 0} = k$. This will be double at time t_d, where

$$\frac{k \cdot e^{rt_d}}{k} = 2 \text{ so } 2 = e^{rt_d} \text{ and } t_d = \frac{\ln(2)}{r} \text{ or } r = \frac{0.7}{t_d}$$

Using specific units, a doubling time of a week corresponds to a rate $r = 0.1$ per day.

COVID-19

That the COVID-19 pandemic was more complicated than described by this simple model is made clear by the graphs of cases and deaths (those with COVID-19 on death certificate) over time, shown for the UK in Figure 4.4.2.

ACTIVITY

Ask small groups to list factors that may have influenced the spread of COVID-19 and the numbers reported to produce the data shown in each of these graphs.

The data, particularly on case numbers, is messy. Indeed, the succession of peaks shown in Figure 4.4.2 looks more like a sequence of different diseases. The complicating factors include

- limited testing in the early stages when the real case numbers were vastly greater,
- uneven patterns of reporting,
- virus mutations, and
- timing of countermeasures used to reduce transmission (masks, lockdowns, etc.).

Restrictions were removed completely in February 2022.

The data on deaths is more reliable, but still affected by the countermeasures, and the varying death rates associated with successive variants, mitigation by vaccination, and improving treatments. Reliable inferences about mechanisms and risks needed more detailed and cleaner data than these.

Figure 4.4.3 shows data from the UK Office of National Statistics (ONS) which measured the level of active infection by testing a random sample of the at-home population in England, each week from July 2021 to June 2022.

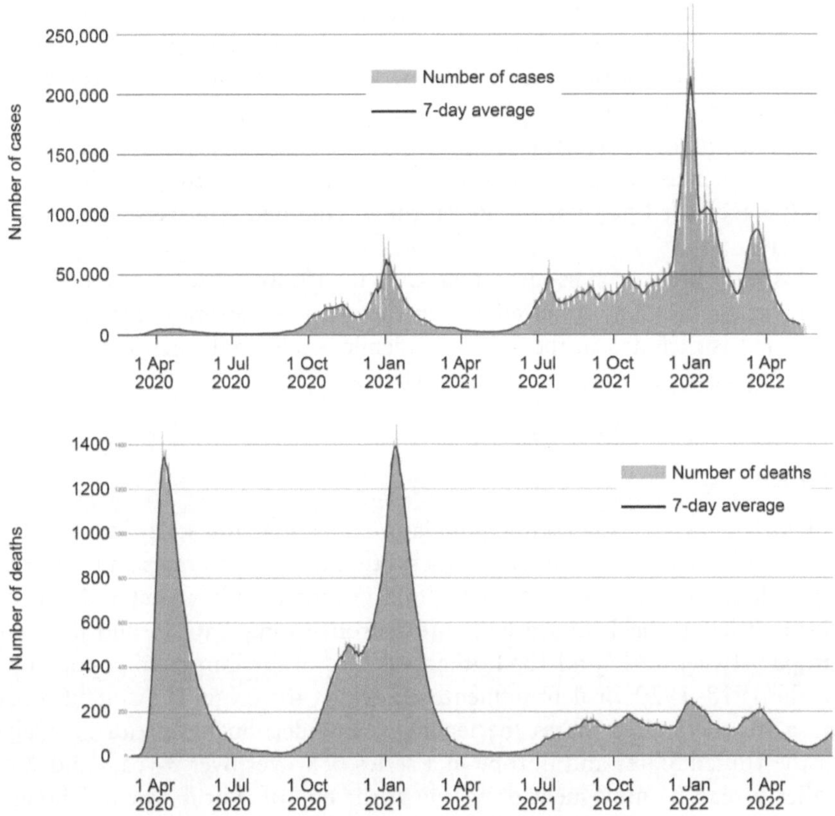

FIGURE 4.4.2 Daily new reported cases and deaths from COVID-19 in the UK, April 2020 to July 2022.

Source: Graphs adapted from coronavirus.data.gov.uk [4G].

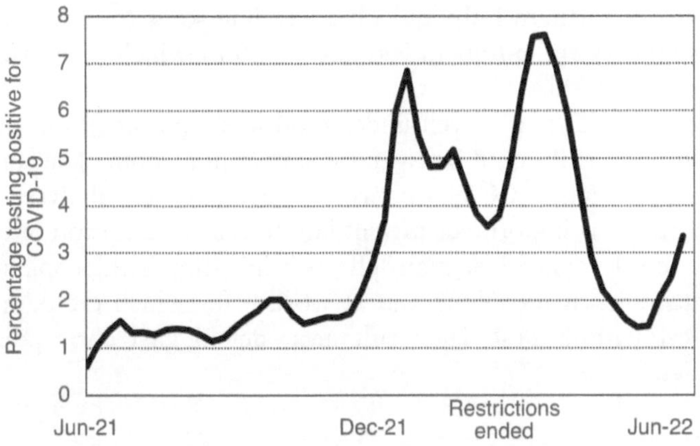

FIGURE 4.4.3 The infection level in England over time for COVID-19.

Source: From weekly random samples – *source data:* ONS Infection survey July 2022 [4G].

It shows much the same qualitative features as the equivalent period in Figure 4.4.2. This data allows a reasonable inference that COVID-19 remained highly infectious, with 1 person in 15–30 having the disease at any one time in the first half of 2022 – as high as at any time in the pandemic. DNA analysis of virus samples showed that successive peaks were associated with increasingly infectious subvariants. Fortunately, with a now largely vaccinated population, death rates remained relatively low in rich countries.

This data alone makes clear that COVID-19, and indeed any real epidemic, brings in many other effects, so the analysis becomes more complex than what is included in the simple epidemic model, with many more variables. No surprise, but what are these effects?

Diffusion

Rather than a closed interacting population, any infection starts in one place and spreads – a complex diffusion process that depends on rates and patterns of people mixing, locally, regionally, and internationally. The highly infectious and lethal Black Death had spread from Mongolia into and through Europe between 1347 and 1351, killing a third of the European population. For the 1918–1920 virulent influenza epidemic, the origin is disputed, with Kansas in the United States the leading contender, but it spread throughout the United States and Europe in a series of waves over 3 years and was further spread by movement of people at the end of World War I. (Though widely known as "the Spanish flu", Spain was certainly *not* the origin – just the first place where the media were allowed to report it.) For recent epidemics like COVID-19, modern air travel greatly speeds up this process. From its first detection in China in late 2019, where it spread rapidly in the city of Wuhan, the COVID-19 virus (properly named SARS-Cov-2) was carried by Chinese skiers to northern Italy, quickly spreading so as to overwhelm the regional hospital system in Italy in February 2020 by which time cases were rising all over the world.

The physics of diffusion is well understood in simple situations like dissolving a solid in a still liquid or cigarette smoke in a room. The spread of virus aerosols in carrying COVID between individuals was analysed at this level, highlighting the importance of ventilation. But the diffusion of the epidemic at a macro-level involved many effects originating with people's habits of mixing at home, in travelling, and at work – for which the parameters were rough estimates at best. The resultant modelling gave correspondingly diverse predictions.

Virus mutation

All organisms mutate through random changes to their DNA; the simplest ones like viruses change more quickly. The process of natural selection ensures that the variants that come to dominate are those that spread fastest (i.e. have higher R_0). Killing people is a side effect that is, in fact, counterproductive from the 'point of view' of a virus that is transmitted when people meet – dead people don't move around. In some severe diseases, like Ebola or SARS-1, people become seriously ill immediately, so R_0 is low, determined largely by the sanitary practices in nursing and burial. The COVID-19 virus was thus a very 'efficient' virus, made more so through its rapid mutation and, crucially, the limited cross-immunity between variants. Thus, the second assumption of the simple epidemic model, that you wouldn't catch COVID twice, was not valid and herd immunity didn't happen naturally.

Variable outcomes with age

For diseases like COVID-19 there are various levels of hazard: death, intensive care, hospitalisation, more or less severe or long-lasting illness at home, or even asymptomatic infection where the person is unaware that they have the disease. For the individual the likelihood of different levels of hazard is crucial for sensible decision-making. Again, we shall focus on risk of dying. We have noted from Figure 4.1.2 that age is the central variable in determining base risk of death.

In Figure 4.4.4, the quantities graphed are base risk of dying and 'COVID IFR', the infection–fatality ratio, calculated here as the percentage of people who get COVID-19 who then die from it. Both graphs in the figure show that, if you catch COVID-19, the age dependence of the risk of dying is remarkably similar to the base risk. This is not always true for epidemics; for example, the 1919 influenza epidemic was particularly lethal for young adults. The absolute risk of a person dying after catching COVID-19 unvaccinated (at that time in the UK) roughly doubled for every 8 years older.

But the figure also shows that the risk of dying if you had COVID was about 60% of your base risk across all adult ages. In the language of Figure 4.4.2, this is a 60% increase in your total micromorts per year, or catching COVID-19 was equivalent to around 7 months (60% of a year) of additional 'normal risk'. So COVID-19 was indeed a serious disease, particularly for older people. As always, more detailed data shows a more complex picture with substantially different population mortality rates for different sub-populations but the overall picture remains much the same.

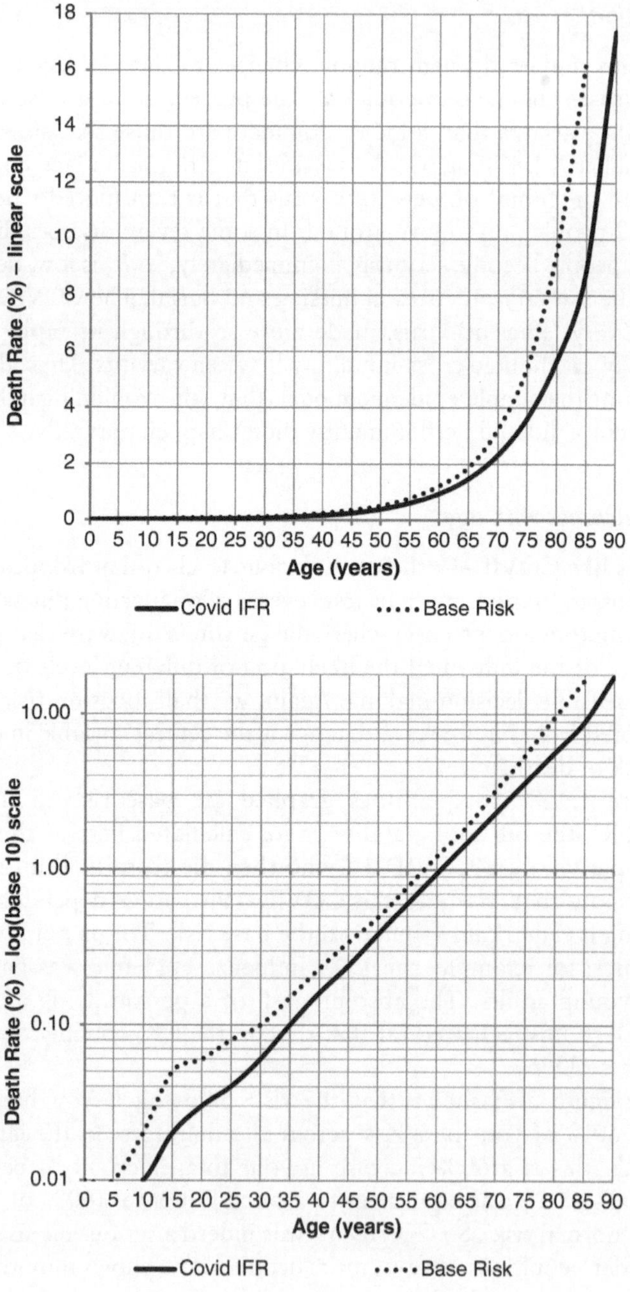

FIGURE 4.4.4 Death rate with COVID-19 and base annual risk versus age on linear (top) and logarithmic (bottom) scales.

Source: UK data through October 2020 (before vaccination became available). Combines data from Brazeau et al. (2020) and ONS life tables.

ACTIVITY

a Discuss the advantages of linear versus logarithmic scales for showing various kinds of functional behaviour – and why these different behaviours arise.

(Ans: Linear versus exponential – because the eye detects straight lines most efficiently)

b Explain with examples how the conclusion that there is a 60% increase due to COVID-19 was reached.

Mitigation measures

The measures taken by governments to slow the spread of COVID-19 covered the whole range. It is true that the spread can be halted if every person is isolated for the time from first infection to becoming non-infectious, about two weeks for COVID-19; but the problems of sustaining life like this are obvious. China nevertheless adopted an approach close to this, with very strict stay-at-home measures; this kept case numbers low until the later more infectious variants 'broke through'. New Zealand used its island isolation to exclude all incoming people, along with a rapid response to the few cases that got through. In the UK there was a series of lockdowns as numbers started to rise – each too late in the view of some experts – with relaxations in between. Influenced by the early pictures from Italian hospitals, the key policy goal was to prevent the health service being overwhelmed – "flattening the peak". In the United States, there was strong political pressure, which varied from state to state, against using restrictions on 'personal freedom' – freedom to infect other people.

Modelling COVID-19

From early 2020 the COVID-19 pandemic led to an intense modelling effort across many teams in many countries to predict the course of the epidemic and to evaluate mitigation options. Ambitious early efforts at prediction produced wildly differing results, but it was then the best guidance available for policymakers in making public health decisions that affected whole populations. In the UK, just before the first lockdown in March 2020, a consensus of the various modelling groups led to an official statement that "20 000 deaths would be a good result"; in fact, total deaths from COVID-19 in the following two years were over 200 000. A bad result.

As time went by and understanding of COVID-19 and its propagation and treatment improved, the modelling became less ambitious and largely descriptive, mainly with more sophisticated statistical analyses of recent data along the lines of the simple epidemic model above, but applied to multiple groups, in different places, with different R and r values. The COVID-19 experience underlines an important general result: that mathematical modelling of complex situations often gives only qualitative insights into the phenomena involved. These can nonetheless be the best information available for individual and community decision-making.

For COVID it was the vaccine research, remarkably accelerated by methodological changes in the trial and approval process, along with improvements in treatment, that produced the breakthrough. Vaccination turned a disease, which if left to run its course might well have increased the normal death rate by more than 50%, into one that was relatively rarely life-threatening. In fact, the rate of infection did not decrease for a long time, driven by variants that were more and more infectious and it is likely to remain significant – an endemic infection. While estimates of R_0 for the original 'wild' COVID-19 virus were around 4.5, estimates for R_0 for the BA5 subvariant of Omicron that emerged in mid-2022 were initially around 18 – matching that for measles (Guerra et al., 2017) along the benchmark for high transmissibility. At least for the vaccinated, the disease caused was much less likely to result in death than early in the pandemic. For further reading, see the article "As COVID deaths in the UK pass the grim milestone of 200 000, what have we learned?" from *The Guardian* [4H].

The overall picture

Perhaps the cleanest data on the effect of COVID-19 on the death rate is the excess deaths over the average for the previous five years, shown by Figure 4.4.5 for England. From April 2020 to June 2022, there were 1.26 million registered deaths against 1.14 million 'expected deaths', which amounts to a 10% increase, most of which could be attributed to COVID-19.

Implications for the individual

In what ways can this picture of the risks of COVID-19 help individuals make better decisions? The data in Figure 4.4.4 shows each of us the absolute risk of dying if you caught an early variant of COVID unvaccinated, and how it relates to your base risk. It is clear that the increase in your base risk is substantial, typically around 60%. As always, how you view this risk is a personal decision. The equivalent data for fully vaccinated people makes it clear that the risk of death (though not of infection) is very much lower. So, unless you have other compelling reasons, you should get vaccinated or take strong measures not to catch the disease. Even for the vaccinated, COVID-19 can be an unpleasant and sometimes dangerous disease, with symptoms that persist for a long time in about 10% of those infected.

The other decisions are mainly about what mitigation measures you choose to take: avoiding spending long periods in crowded indoor places, wearing a

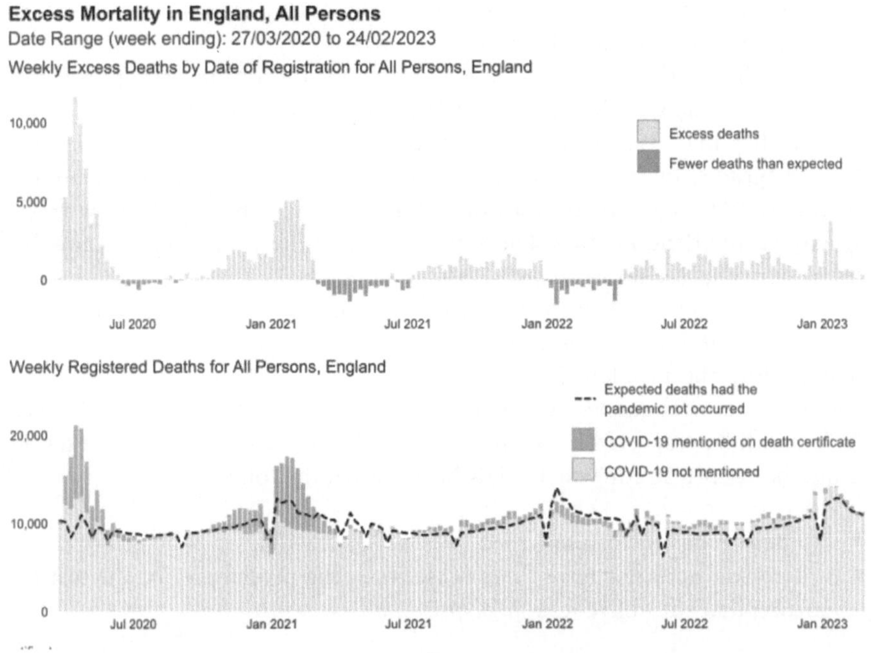

FIGURE 4.4.5 Excess deaths in England during COVID-19 from March 2020 to February 2023.

Source: Graph adapted from UK Office for Health Improvement and Disparities [4J].

mask, particularly when in such places, having your groceries delivered? For such questions, finding and analysing the data provides a quantitative basis on which the mathematically literate person can make informed decisions.

A readable analysis of the UK COVID data and our understanding of how the epidemic developed is given in the book *COVID by Numbers* by Spiegelhalter and Masters (2021).

4.5 Medical testing

Medical testing for infections and other 'morbidities' is an important part of healthcare, in which improvements over recent decades have enabled earlier treatments and improved outcomes for many serious diseases. But no test is perfect – each test gives proportions of 'false positive' and 'false negative' results. This makes it an interesting and motivating context for developing an understanding of conditional probability – a notoriously tricky topic – in a risk-related context. This is partly because it produces some surprising results; most of us find it surprising at first that even with a test that gives a small proportion of false positives, most of the positive test results can be false. For the mathematically interested person, surprising results like this are a reason to look more deeply.

Teachers and their students take a variety of approaches to representing the situation: using probabilities or calculating numbers within a given population. They may use various representations: tree diagrams, tables, or even Venn diagrams. In any approach, stating exactly what is being calculated is crucial – and tricky: "the proportion of false positives" is ambiguous without specifying the denominator. The standard use is the proportion of positive test results that are false.

The Mathematics Assessment Project's 'classroom challenge' on *Representing Probabilities: Medical Testing* [4K] provides a sequence of activities that explore the problem in some depth. It starts with a specific problem, comparing two countries A and B with different incidence of the disease, 20% and 2%, respectively. Samples of people from each country are tested using the same test, for which trials have shown the following:

- If a person has the disease, then the test result will always be positive.
- If a person does not have the disease, then the probability of the test reporting that they have it, a '*false positive' test result*, is 5%.

The question posed, for each country, is as follows:

ACTIVITY

Suppose a patient is told that they have tested positive. What is the probability that the test is wrong?

The tree diagram representation for country A is shown in Figure 4.5.1. It shows that the percentage of positive tests that are wrong is

$$\frac{0.04}{0.2+0.04}=17\%$$

For Country B, with 2% incidence, the corresponding numbers are shown in Figure 4.5.2. The 71% of positive tests that are false in this case arises because the true positives drop in proportion to disease incidence; the false positives increase only slightly. This is a nice example of how mathematical literacy enables us to understand surprising results.

The spreadsheet shown in Figure 4.5.2 (see [4L]) enables students to explore the effect of changing the input parameters. It also covers a test that has a false negative rate as well. Changing the population entry from 1 translates probabilities into numbers in a population throughout the calculation. Perhaps most usefully, the 'Show formulas' mode of the spreadsheet brings out the detailed reasoning, which many students find difficult to construct.

In the MAP unit the students individually try the problem, which has some guidance from specific questions, in a previous lesson. A hint sheet is provided for the teacher to use where needed – differentiation through support,

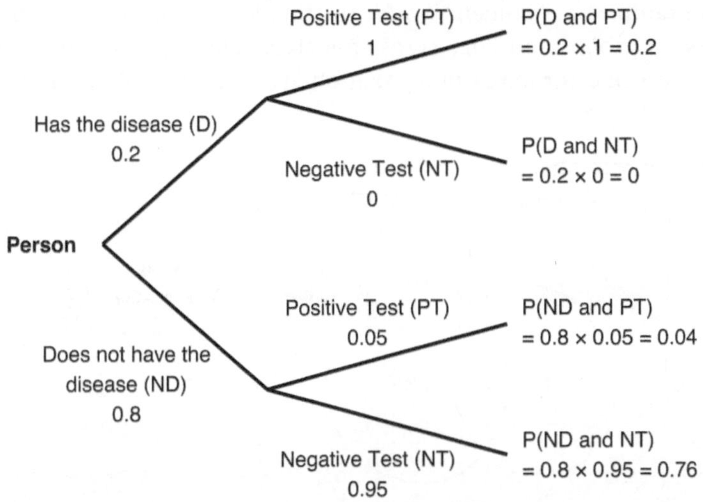

FIGURE 4.5.1 Probabilities for Country A with 20% incidence and a 5% false positive test.

Source: Adapted from *Representing Probabilities: Medical Testing* [4K].

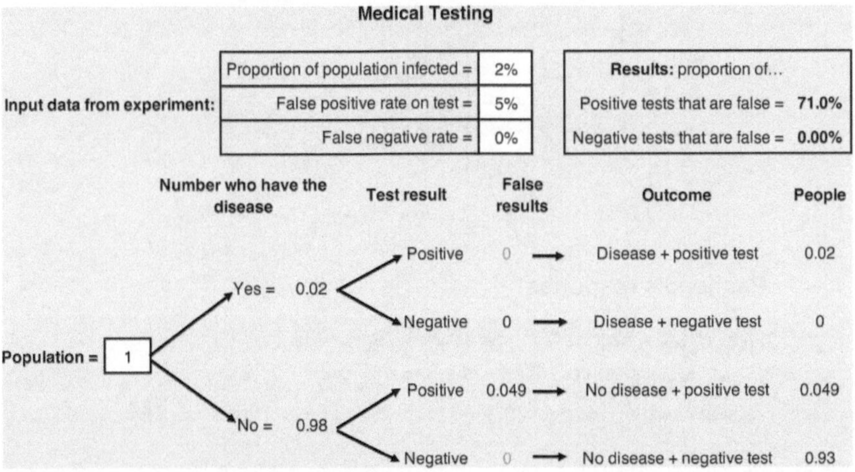

FIGURE 4.5.2 Probabilities for Country B with 2% incidence and a 5% false positive test.

Source: Spreadsheet available at [4L].

not in the core task, is a design principle. Looking through the responses gives the teacher a base for formative assessment.

In the main lesson, the teacher clarifies the definition of false positives, supported by an applet. Students share their attempts in small groups, deciding on a common approach. The focus is on the chains of reasoning. In the diagnostic teaching approach to formative assessment, a key design tactic uses examples of "student work" to stimulate discussion of various approaches, in groups and later across the class. The examples are chosen to show sources of error in reasoning that often arise. Figure 4.5.3 shows three.

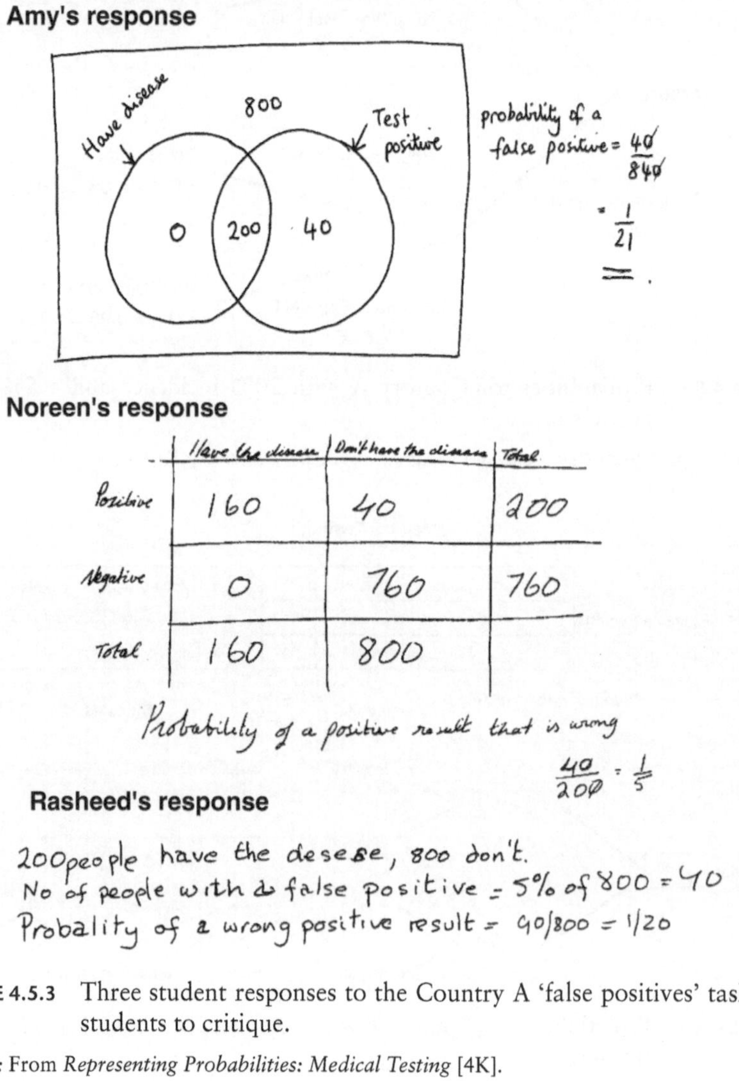

FIGURE 4.5.3 Three student responses to the Country A 'false positives' task for students to critique.

Source: From *Representing Probabilities: Medical Testing* [4K].

These use the idea of 'expected frequencies', for example, what does it mean for 1,000 people being tested? This is clearer to most people than the (equivalent) probability tree in Figures 4.5.1 and 4.5.2.

Amy has an error in her table – there should be 200 people, not 160, who have the disease and test positive – but her method, for the proportion of positive tests that are false, is correct. Noreen and Rajeev calculate a different quantity – the proportion of false positives in the whole population. Rajeev's calculation is correct, and well explained. Noreen has an error in her Venn diagram – there should be 760 people, not 800, who do not have the disease and test negative.

This device leads to focused discussion, moving students into the 'teacher roles' of critiquing and explaining – a role shift that, as explained in Chapter 2, raises the level of discussion from answers to reasoning. No arithmetic errors are involved – trials showed that they divert attention from the reasoning.

Acknowledgements

Our thanks to David Spiegelhalter for educative and entertaining conversations on risk over many years.

Much of the data used in this chapter comes from UK government sources – such as the Office for National Statistics and the Office for Health Improvement and Disparities – and is public sector information licensed under the Open Government Licence v3.0. See the individual attributions and links for more information.

Material from the two versions of the *How risky is life* materials appears courtesy of the *Bowland Maths* maintainers and the Australian Academy of Science. Specifically, Figures 4.2.1 and 4.2.2 are from *Bowland Maths* © Bowland Charitable Trust 2008. Bowland Maths materials are free for educational use. Figure 4.2.3 is from *reSolve: Mathematics by Inquiry* © Australian Academy of Science, available under Creative Commons CC-BY-NC-SA licence. Similar material appears in both versions. The software shown in Figure 4.2.4, which features in both *Bowland* and *reSolve*, is from the Shell Centre for Mathematical Education and available under Creative Commons CC-BY-NC-SA licence.

Figures 4.5.1 and 4.5.3 adapted here with permission from the Bell Burkhardt Daro Shell Centre (BBDSC) Trust on behalf of the Mathematics Assessment Project. © Mathematics Assessment Resource Service 2007–2015. The *Classroom Challenges* materials are available under the Creative Commons BY-NC-ND 3.0 licence.

References

Brazeau, N., Verity, R., & Jenks, S. et al. (2020). COVID-19 infection fatality ratio: Estimates from seroprevalence. Imperial College London. https://doi.org/10.25561/83545

Guerra, F. M., Bolotin, S., & Lim, G. et. al. (2017). The basic reproduction number (R_0) of measles: A systematic review. *Lancet* 17(12). https://doi.org/10.1016/S1473-3099(17)30307-9

Hoffrage, U., Lindsey, S., Hertwig, R., & Gigerenzer, G. (2000). Communicating statistical information. *Science* 290, 2261–2262. https://doi.org/10.1126/science.290.5500.2261

Howard, R. A. (1980). On making life and death decisions. In J. Richard, C. Schwing, & W. A. Albers (Eds.), *Societal risk assessment: How safe is safe enough?* General Motors Research Laboratories. Plenum Press.

Spiegelhalter, D., & Masters, A. (2021). *COVID by numbers: Making sense of the pandemic with data*. Penguin UK.

Links to useful material

To visit any of these links, scan this QR code or visit ltml.mathlit.org – append the link code to go directly to the entry – for example, ltml.mathlit.org/4A

The original source links are given below for attribution purposes:

[4A] *National Life Tables* – Office for National Statistics (UK)
https://www.ons.gov.uk/peoplepopulationandcommunity/birthsdeathsandmarriages/lifeexpectancies/datasets/nationallifetablesunitedkingdomreferencetables

[4B] *How Risky Is Life?* – Bowland Maths & reSolve
https://www.bowlandmaths.org /projects/how_risky_is_life.html
https://www.resolve.edu.au/mathematical-modelling-how-risky-life?lesson=1672

[4C] *Visualising Small Risks* – Bowland Maths & reSolve
https://ltml.mathlit.org/4C

[4D] *Drowning Reports* –Royal Life Saving Australia
https://www.royallifesaving.com.au/research-and-policy/drowning-research/national-drowning-reports

[4E] *Simulating Random Variations* – Shell Centre for Mathematics Education
https://ltml.mathshell.org/4E

[4F] *Basic Epidemic Model* spreadsheet
https://ltml.mathlit.org/4F

[4G] *UK COVID-19 data sources* – UK government & UK Office for National Statistics
https://coronavirus.data.gov.uk/
https://www.ons.gov.uk/peoplepopulationandcommunity/healthandsocialcare/conditionsanddiseases

[4H] *As COVID Deaths in the UK Pass the Grim Milestone of 200,000, What Have We Learned?* – The Guardian
https://www.theguardian.com/commentisfree/2022/jul/13/as-covid-deaths-in-the-uk-surpass-the-grim-milestone-of-200000-what-have-we-learned

[4J] *Excess Mortality in England* – Office for Health Improvement and Disparities
https://www.gov.uk/government/publications/excess-mortality-in-england-weekly-reports

[4K] *Representing Probabilities: Medical Testing* – Mathematics Assessment Project
https://www.map.mathshell.org/lessons.php?unit=9405&collection=8

[4L] *Medical Testing* spreadsheet
https://ltml.mathlit.org/4L

[4M] *Released Mathematics Items*– PISA/OECD
Test questions 2006 https://www.oecd.org/pisa/testquestions-pisa 2006.htmTest questions 2003 – 2012 https://www.oecd.org/pisa/pisaproducts/pisa-test-questions.htm
Direct link: https://www.oecd.org/pisa/test/PISA%202012%20items %20for%20release_ENGLISH.pdf

5

CLIMATE CHANGE

The science, the mathematics, and the politics

In the long term there is no greater risk than climate change, or other topic as important – hence this separate chapter where, as usual, we hope to set out some of the basic underlying phenomena, physical and societal, in a way that stimulates constructive discussions between students on a subject that is of interest and concern to so many of them. Because of the complexity of these systems, this is mainly a 'know about' chapter in a field where modelling and data complement each other – but the models are complex.

The effects of human activity on the climate, predicted for over 50 years, are already with us – in droughts, disappearing glaciers, and the worldwide increase in extreme weather events, to name just a few examples. The dynamics of the climate and its impact on weather are complex but the basic science is clear, as is the kind of changes in human behaviour needed to mitigate the effects of the heating.

In this chapter we look at how mathematical literacy can help each of us to better understand the evidence on climate change, and the underlying dynamics. We go on to look at the societal challenges and the actions, personal and political, that will mitigate the damage that is ongoing – and possibly even reverse some of it. We look at data, the pattern of arguments that are deployed, some of the models being presented and the questions they raise.

After first looking at the science we go on to describe the political and social frameworks within which climate change has been discussed since it was first *widely* recognised by scientists as a potential problem in the 1970s. This is key to relating an apparently straightforward set of scientific results to the range of human responses to these results.

DOI: 10.4324/9781003303503-6

Although relatively modest changes in climate can cause – and are already causing – problems, a greater concern is that continued inaction could create a self-reinforcing cycle of escalating change with even more catastrophic results. So we look in more detail at the mathematics of feedback and tipping points which makes such concerns credible.

Finally, we point to the parallel linked challenges of understanding the much more complex effects of human activity on the ecosystem of the planet and the methodological and ethical issues they raise.

5.1 The science of global warming

Variations in climate have been happening for millions of years, some substantial ones within timescales of tens of thousands of years. The last Ice Age lasted about 100 000 years, ending about 10 000 years ago. The global average temperature was about 5°C lower than today. Larger variations happen in specific places from year to year and day to day – these we call *weather*. The current focus is on changes with a timescale measured in decades and how far these are *anthropogenic*, that is caused by human activity.

An awareness that changes in atmospheric carbon dioxide levels could substantially alter the surface temperature through 'the greenhouse effect' goes back over 100 years to a paper by the Swedish scientist Svante Arrhenius (1896). In 1938 Guy Callendar connected carbon dioxide *increases* in Earth's atmosphere to observed global warming. By the 1960s internal CIA reports recognised the phenomena and began to estimate its effects, societal as well as physical. But the 1970s was the starting point of serious *political* interest. In 1972 John Sawyer, a scientist at the UK Meteorological Office, reviewed the scientific knowledge at the time on the likely effects of anthropogenic carbon dioxide as a 'greenhouse gas', its distribution and rise. His article (Sawyer, 1972) stated:

> The increase of 25% CO_2 expected by the end of the century therefore corresponds to an increase of 0.6°C in the world temperature – an amount somewhat greater than the climatic variation of recent centuries.
>
> *(p. 25)*

In the outcome Sawyer rather accurately predicted the rate of global warming for the period between 1972 and 2000, shown by the subsequent measurements from three national centres, such as the NASA data in Figure 5.1.1.

This data begins in the nineteenth century, when thermometers were introduced into meteorology. But we can now measure temperatures, and many other variables, from thousands of years ago. How? One answer

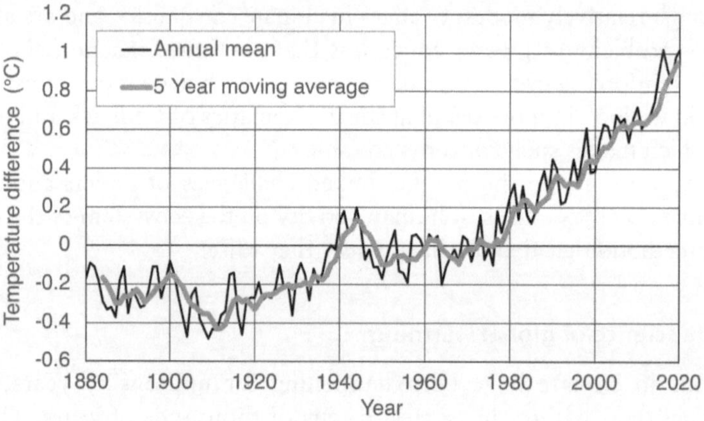

FIGURE 5.1.1 World average mean annual temperature relative to the average for 1951–1980.

Data source: NASA's Goddard Institute for Space Studies (GISS) [5A].

lies in an ingenious bit of science and mathematics involving 'ice cores'. These are vertical columns of ice cut from the Antarctic ice covering, which is formed from snow that has fallen over centuries, each layer compressing those below. Some cores go up to 3 km deep representing 800 000 years of atmospheric history. The physical properties of the ice and of material trapped in it tell us about the climate over the age range of the core. The proportions of different oxygen and hydrogen isotopes provide information about ancient temperatures, and the air trapped in tiny bubbles can be analysed to determine the level of atmospheric gases such as carbon dioxide, where carbon dating (discussed below) is a powerful tool.

The data from both thermometers and ice cores shows that the mean temperature at the Earth's surface has increased by about 1.2°C in the last 150 years – essentially since the industrial revolution literally 'gathered steam'. There are some other sources of data on past climate. On a geological timescale, the pattern of life revealed in the fossil record gives some indication. Over the last several hundred years, tree ring sequences give information on local temperature and rainfall.

5.2 The mechanisms of global heating

Let us begin with some basic scientific facts:

- Carbon dioxide (CO_2) is produced in great quantities by burning current or fossilised organic matter that is rich in carbon: wood, coal, oil, or 'natural' gas. The atmosphere provides the oxygen.

- CO_2 traps the energy of sunlight within the atmosphere, as the glass traps it in a greenhouse. The visible light of the sun passes through the 'glass' and warms the earth; the low temperature infrared light emitted by the warmed earth (invisible to the eye, but you can sometimes feel it) cannot penetrate the atmosphere to escape back into space. Some other gases also contribute to the 'glass'.
- CO_2 is inert and stays in the atmosphere for a long time, with a 'half-life' of about 120 years – that is, half of what is present disappears each 120 years.
- Natural processes create CO_2 – animals breathing out, for example – but this is part of the balanced 'carbon cycle' in which plants *absorb* CO_2 from the air, animals feed (directly or indirectly) on those plants and release some of that recently absorbed CO_2 back into the air. CO_2 released by burning fossil fuels, on the other hand, comes from carbon that has been locked away from the environment for millions of years, and is the main cause of *increasing* CO_2 levels.

That CO_2 absorbs the infrared (thus preventing it radiating into space) is straightforward to demonstrate. Figure 5.2.1 shows one experiment (Levendis et al., 2020) [5B]. A heating element at 50°C is hung in the middle of a 'bouncy ball' balloon filled with either CO_2 or air. After being turned off, the heating element cools more slowly with the CO_2, an effect that increases with increasing amount of CO_2.

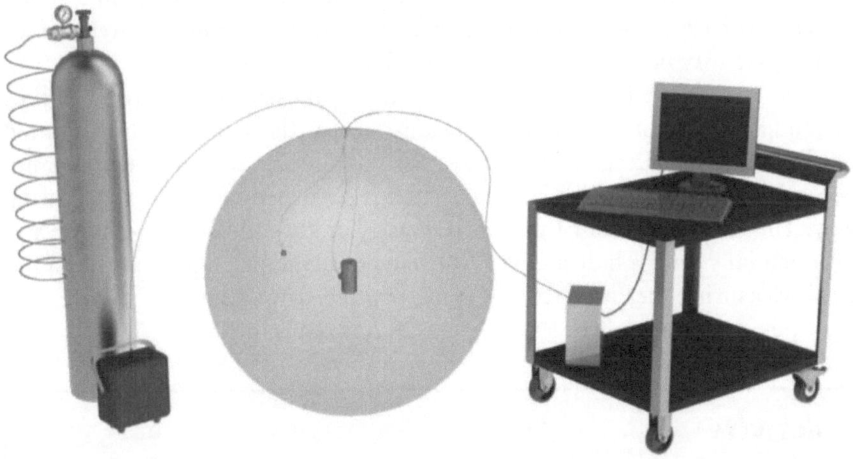

FIGURE 5.2.1 Schematic of an experiment to show the infrared absorption by CO_2.

Source: From Levendis et al. (2020) [5B].

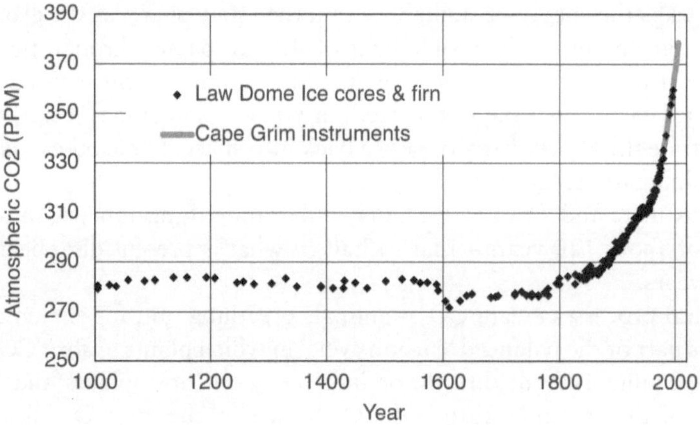

FIGURE 5.2.2 Concentration of CO_2 in the atmosphere— over last 1000 years.
Data source: MacFarling Meure et al. (2006) via NOAA.

That burning carbon produces CO_2 is well known (see, for example, [5C]). One tonne of coal produces about 2.4 tonnes of CO_2; not quite the ratio you would expect from the molecular weights ($C = 12$, $O = 16$, $CO_2 = 44$) because coal is not just carbon.

The atmospheric effect of burning carbon in its various forms is shown in Figure 5.2.2 which shows the parts per million of CO_2 in the atmosphere over the last 1000 years. The line shows data from recent direct measurements in the atmosphere. The blobs are data from ice cores collected by the Australian Scientific Expedition to the Antarctic, from the Law Dome which rises to over 1900 meters above sea level. (*Note:* 'Firn' refers to a type of ice that is part-way between snow and glacial ice.)

Each layer of ice in the core can be dated by measuring the proportion of the radioactive isotope ^{14}C that remains. How does this 'carbon dating' work? ^{14}C is produced in the atmosphere mainly by cosmic neutrons and, because it is chemically the same as ^{12}C, it is absorbed by plants in the usual way. This produces about 1 part in 10^{12} of ^{14}C in the atmosphere. This decays exponentially with a half-life of 5700 years to make ^{12}C, and emitting radiation. Measuring the radioactivity being emitted from a layer in the ice core indicates how much of the ^{14}C is left, and thus dates the layer.

ACTIVITY

Find other uses of carbon dating and how they work – in archaeology and in other fields. How does carbon dating link to tree ring dating?

Further confirmatory evidence from ice cores showed that the proportion of CO_2 and temperature had gone up and down together in wide swings through past ice ages. This confirmed the planet-scale CO_2–temperature relationship in a manner entirely independent of computer climate models, strongly reinforcing the emerging scientific consensus. The findings also pointed to powerful biological and geochemical feedbacks – we discuss feedback mechanisms below.

Looking more deeply

There are other gases that contribute to the greenhouse effect but they have far shorter half-lives, so they disappear more quickly:

- Methane, which is released from oil wells, fracking, and the burping of cows, is 80 times more absorbing than CO_2 but with a half-life of 'only' 10.5 years.
- CFCs, chlorofluorocarbon molecules which have been used in refrigerators, have half-lives from 16 to more than 500 years. These were mainly responsible for the 'hole' in the Antarctic ozone layer that helps protect us from the ultraviolet in sunlight. Thanks to a worldwide effort to eliminate the most damaging CFCs, the ozone layer has recovered.
- Nitrous oxide, which has a half-life of 132 years, also contributed to ozone layer depletion.

But none of these other gases are emitted in anything like the same quantity as CO_2. Global carbon dioxide emissions from fossil fuels and other industrial processes are at record high levels of about 40 billion metric tonnes ($GtCO_2$) per year, about 5 tonnes per person on Earth – the US and European average is several times that. About 2% of this is produced by humans breathing out – about 100 kg per person per year. To measure the differential effects of the various greenhouse gases, the IPCC introduced the concept of 'global warming potential', comparing the effects of equal masses of greenhouse gases over different periods of time, taking into account their half-lives and infrared absorption. CO_2 is used as the reference gas, so its GWP is set at 1 for any time period. Some 100-year GWPs are shown in Table 5.2.3. Most gases with very high GWP have only a low concentration in the atmosphere, so they are not the main factor contributing to climate change. Finally, there is water vapour, a most powerful greenhouse gas because of its infrared absorption. Ironically, it has negligible GWP because human-generated water vapour (e.g. from power station cooling towers) leaves the atmosphere within a week or two. However, it plays a crucial role in the feedback mechanisms that we discuss below.

TABLE 5.2.3 Global warming potential of some greenhouse gases

Greenhouse gas	*Chemical formula*	*Half-life years*	*100-year GWP*
Carbon dioxide	CO_2	5700	1
Methane	CH_4	12	25
Nitrous oxide	N_2O	109	298
Water vapour	H_2O	~0	~0
Some fluorocarbons – refrigerator gases			
HFC-23	CHF_3	222	14 800
Difluoromethane (HFC-32)	CH_2F_2	5	675
Fluoromethane (HFC-41)	CH_3F	2	92

Source: Compiled using data from Wikipedia [5D].

ACTIVITY

Using this and other data as necessary, estimate the 25-year and 500-year GWPs of these gases.

5.3 The politics of climate change

The 1970s was the starting point of serious political interest in the effect of CO_2 emissions on the climate. It perhaps began with John Sawyer's 1972 prediction of an increase of 0.6°C in the world temperature by 2000. By 1977 Frank Press, US President Carter's Science Adviser, had sent the president a memo stating:

> Because of the "greenhouse effect" of atmospheric CO_2 the increased concentration will induce a global climatic warming of anywhere from 0.5 to 5°C. ... The potential effect on the environment of a climatic fluctuation of such rapidity could be catastrophic and calls for an impact assessment of unprecedented importance and difficulty. A rapid climatic change may result in large scale crop failures at a time when an increased world population taxes agriculture to the limits of productivity. ... The urgency of the problem derives from our inability to shift rapidly to non-fossil fuel sources once the climatic effects become evident not long after the year 2000.

Around the same time, internal research at the US oil companies came to much the same conclusions. Supran et al. (2023) [5E] show that one of those fossil fuel companies, ExxonMobil, had their own internal models that projected warming trajectories. Figure 5.3.1 compares Exxon's predictions with

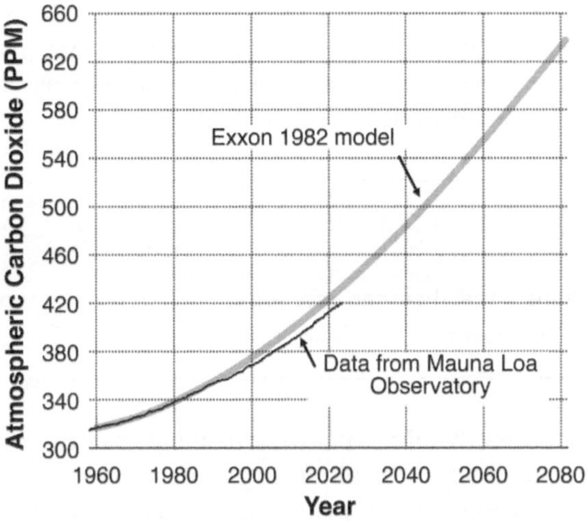

FIGURE 5.3.1 Atmospheric concentration of CO_2 – predicted versus measured.

Source: Data from ExxonMobil 1980s predictions (Supran et al., 2023) [5E] and NOAA measurements [5F].

modern measurements from NOAA/NASA. Their predictions were consistent with those forecast by the independent academic and government models, and were confirmed by subsequent observation of atmospheric CO_2 (*NOAA via NASA,* [5F]).

Yet for decades members of the fossil fuel industry have worked to convince the public that anthropogenic climate warming is not real or, at least, not serious. What they knew about climate change thus contradicted what they led the public to believe. The obvious commercial incentive to minimise changes in the balance of different sources of energy is pursued through a variety of strategies like those developed by the tobacco industry in the 1950s, as evidence grew that smoking causes lung cancer. They argue that the models used are too uncertain to show a causative link between the increase in CO_2 from fossil fuel use and climate, and that the clear correlation does not imply causation. Yet, for climate change as for cancer, the predictions of the models have proved remarkably accurate over half a century and causation has been established through identifying the mechanisms underlying the observations. Prediction and then testing against subsequent events, and understanding mechanisms are fundamental in establishing scientific validity.

Gradually over time there has been a shift in the defensive arguments towards accepting global warming is a fact but denying its seriousness and so the need to do much about it. This is often combined with 'greenwashing' – publicity from oil companies and others that aim to improve their image by describing steps to reduce carbon emissions that exaggerate their likely effect.

For example, 'carbon offsetting', where trees are planted that will, over decades, absorb as much CO_2 as was emitted in the flight they just took yesterday. There can never be enough trees in the world to offset current carbon emissions; there is no substitute for cutting emissions.

ACTIVITY

Justify the statement that there can never be enough trees in the world to offset current carbon emissions.

The 'debate' continues to shift. The overwhelming majority of scientific papers (literature reviews find 97% or more) provide evidence of anthropogenic causes of climate change and its potentially serious consequences: sea level rises causing floods of coastal land, some places becoming uninhabitable through temperature, food crop failures, and mass migrations. The fossil fuel companies continue to construct specious arguments, finding – and funding – scientists who support them. They are not alone. There are also strong strands of political and public opinion that seek to undermine both the scientific facts and, particularly, the need for making challenging societal changes. Unfortunately, the current tendency to polarisation of political beliefs is making resolution through rational argument based on data more difficult.

There are two main dimensions in the way governments and their citizens have responded: danger level and action planned. The range of response in each is broad, according to the assessment of risk.

Danger level:

1 *Hoax:* humans have no significant influence on climate.
2 *Minimal effects:* there are some effects but nothing to worry about.
3 *Serious challenges:* this is a problem and we must adjust our lives to address it.
4 *Potentially catastrophic:* this is an upcoming disaster which must be top priority in policymaking and in individual choices.

Action required:

a None.
b Awareness, along with some personal efforts to reduce our CO_2 profile.
c Modest political adjustment, with policies that claim to meet international targets – but, given how far rhetoric is ahead of action, they are unlikely to do so.
d Steps that amount to a major change now in the way we live.

Although we lack data, it seems likely that in many Western countries there is a significant proportion of 1a people, a majority in 3b or 3c, and a small proportion of really concerned 4d – along with many 'don't know's. As we write most governments seem to take a 3c position, 3d or 4d being politically challenging for all governments:

- For democracies because of the need to keep voters on side.
- For autocratic regimes because of a general priority for economic growth.

The mass of data available for thinking through these issues presents both an opportunity and a challenge in trying to come to a reasoned position.

ACTIVITY

Working in small groups, gather and analyse data to prepare an argument for a debate on what actions should be taken to reduce global warming, and to mitigate the effects of climate change. Each group will take a specific role, for example, fossil fuel company publicist, climate change activist, conservative or progressive politicians, leader of a Pacific Island, marine biologist, world food or human rights advocate. (Draw lots if necessary to ensure coverage.)

5.4 The mathematics of feedback and tipping points

Because we have been setting out something of the complexities of climate change, and the scientific understanding of it, most of the mathematical literacy demands have been in been understanding the meaning of data, presented in numbers or graphs, and the units and rates involved and proportional reasoning about its implications.

We conclude this chapter with the mathematics of two dynamical mechanisms that can make change catastrophic: feedback and tipping points. They are found in many contexts, including aspects of climate change. Both can be modelled within standard mathematics in ways that vividly illustrate the general principles.

Feedback

Feedback (strictly 'positive feedback') is the term used when a change leads to more of the same change. For example, the white ice in the Arctic Ocean reflects a lot of sunlight back into space, whereas open water absorbs it and warms up. This creates a positive feedback loop:

1 Temperatures of ocean and atmosphere increase slightly (for any reason).
2 → There is less ice and more open water.
3 → Less sunlight gets reflected and more is absorbed by the ocean.
4 → The temperature of the ocean (and then the atmosphere) increases further.
5 Repeat from step 2!

To take a more mathematical approach, say that the ice cover in the Arctic Ocean in the winter lasts for a period T each year. If the warming effect comes largely from the planet as a whole – that is, there is no "feedback" from local changes – and we assume it is steady over time, there will be a linear decline in T from the starting year T_0 to the year when $T = 0$. (Why does the model stop working when $T = 0$?)

If, however, the *local* warming effect is also important, then the *rate* of reduction will increase as T gets shorter, so the solutions to the new models show exponential behaviour, characteristic of feedback. It cannot go on forever, of course, as other mechanisms intervene.

$\frac{dT}{dt} = -a(T_0 - T)$	where t is the elapsed time and a is assumed constant
$-\ln(T_0 - T) = -at + b$	solving the differential equation, with b a constant
$T_0 - T = c.e^{at}$	where constant b is $-\ln(c)$
$T = T_0 - c.e^{at}$	the ice cover reduces exponentially over time

In reality, both local and global effects are present which lead to a more complicated model, but still expressible and solvable by calculus. As usual, when the model is further improved by taking other factors into account, numerical methods are required. Numerical methods, step by step using a spreadsheet, can be used to demonstrate the effect to pre-calculus students.

There are many feedback effects in the dynamics of climate. The most important ever-present one is more complex: the effect of water vapour. This arises because the amount of water that the atmosphere can hold increases rapidly with temperature. This drives the observed and predicted increase in storms and other extreme weather events. Water vapour amplifies the effect of CO_2; without it, even doubling the pre-industrial level of CO_2 (to 560 ppm) would lead to an increase in global temperature of only about 1°C. The sea is also a principal repository of CO_2; the amount it can hold decreases with rising temperature, creating another positive feedback mechanism.

Feedback mechanisms make modelling the climate with mathematics very challenging. Although the many simulations of the diverse modelling groups vary a lot in detail, the overall picture is consistent.

Tipping points – obvious and surprising

A tipping point arises when an infinitesimal change in the parameters of a system produces a finite shift in the state of the system.

Tipping points are a familiar phenomenon. In the literal sense of 'tipping point', a tall object will topple over if it is tilted just a tiny bit beyond a certain limit. Generalising the concept, when an election is lost by even a few votes, a government can change. A kettle boils when the temperature reaches 100°C – then it stays there, all the extra energy making steam. The popularity of soccer is said to arise from the tension that arises because each goal is seen as a potential tipping point in the match. The claim is that the game is fascinating *because* there are so few goals not as some might think, despite that.

How might tipping points affect climate change? There are various systems of concern. The ground in a huge area of Siberia is frozen throughout the year. This *permafrost* holds a large amount of CO_2 which will be released into the atmosphere if the ground temperature moves above 0°C. The temperature is rising fast in Siberia, hence the concern about this tipping point. The Atlantic Meridional Overturning Circulation (AMOC) is a system of ocean currents, including the Gulf Stream, that redistributes heat by bringing it from the tropics all the way to Greenland and carrying cold water back to the equator; it is weakening and, if it reaches a possible tipping point, it could bring back Ice Age conditions to Europe. A very surprising outcome of global warming.

How does mathematics model tipping points? Let's start with some literal tipping points. When a ladder leans against a wall, there are at least six ways it can tip over; two are sketched in Figure 5.4.1.

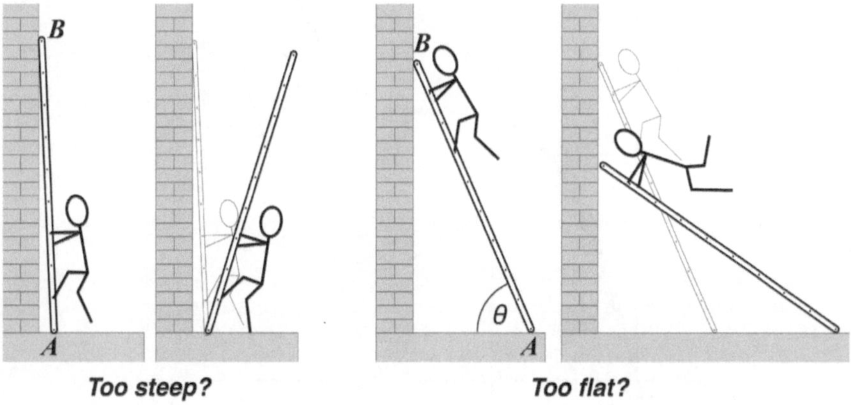

FIGURE 5.4.1 Two potential tipping points for ladders.

The "too steep" instability shown in the Figure is obvious, and benign. As you step onto the ladder, your weight, acting through your hands, pulls the ladder backwards. It begins to tip away from the wall, so you step off and set it at a less steep angle. The "too flat" case has long been used as a standard high school mechanics task – if the angle of the ladder is too flat, the bottom slips away because the friction at *A* is inadequate. The student has to relate this critical angle θ to the coefficient of friction. Ironically, this *is* the most dangerous of many possible instabilities because it gets worse as the climber goes up the ladder; once the climber is past the centre of the ladder, the ladder is more likely to slip than before – but there is no climber in the traditional examination task! In the real world, this instability explains the traditional safety rule about having another person standing on the bottom of the ladder throughout, so the centre of gravity of the pair remains below that of the ladder itself – provided they are of comparable weight, not Laurel and Hardy.

The mathematical analysis of tipping points was called Catastrophe Theory by René Thom, who won a Fields Medal ("The equivalent in Mathematics of a Nobel Prize") for classifying topological features of surfaces that produce tipping points (Thom, 1977). We can get the idea from a simple one-dimensional case shown in Figure 5.4.2.

When $P = 0.5$, the 'ball' in Figure 5.4.2 is sitting 'safely' in the dip of this helter-skelter cubic curve – the nearby point of lowest gravitational potential. Even if the ball is displaced slightly, it will roll back to the stable position. The other curves show the graphs for various values of P. Note how the dip gets shallower as P increases, disappearing at $P = 4/3$ – the tipping point when,

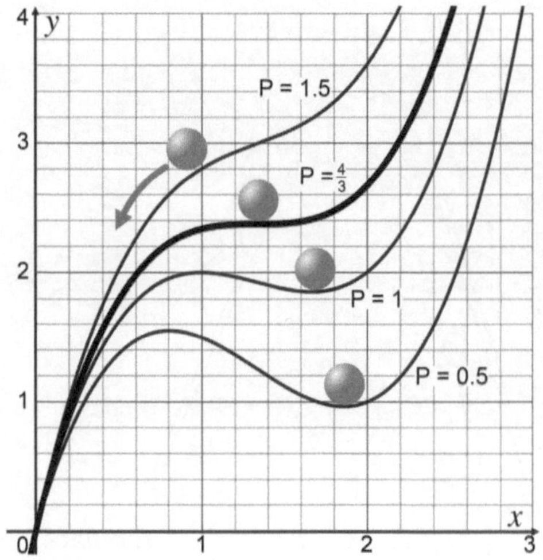

FIGURE 5.4.2 Graphs of the potential function $y = x\left(P + (x-2)^2\right)$.

mathematically, the two roots for $dy/dx = 0$ coincide, so the maximum and minimum merge into a point of inflexion. There the curve is completely flat. Physically, with no local minimum potential, there is nothing keeping the ball in position, so the slightest displacement from the inflection point will send it rolling away down the slope. With $P > 4/3$, there is no flat area at all for the ball to rest.

Even with $P = 0.5$, push the ball too far to the left and it will roll down the slope. As P approaches the 'tipping point' of 4/3, the 'safety margin' – how far the ball can be moved and still return to the equilibrium point – becomes smaller. This is an example of a 'metastable equilibrium' which is a common feature of the cycles and balances on which our environment depends. Artificially pushing the environment close to a tipping point reduces its resilience to other, unavoidable disturbances with natural causes.

The phenomena described in this chapter – physical, mathematical, and societal – underpin the need for critical thinking that leads to *far more effective action* than the start that societies around the world have made so far.

5.5 Climate change and the environment

We cannot leave the issue of human activity and climate change without referring to their direct consequences for the ecosystem in which we live. The environmental effects of climate change are even more complex and diverse than the physical changes we have discussed so far – and correspondingly more difficult to model. Sorting and understanding the data alone is a challenge, even before tangling with issues of causality. Here we shall outline some smaller scale phenomena which illustrate these challenges before going on to discuss some theoretical frameworks people have put forward to think about ecological impacts.

Climate versus environment – some examples of complex trade-offs

Efforts to reduce climate change can bring their own environmental concerns. The following are examples of the unintended consequences of well-intentioned initiatives. They illustrate the difficulty of designing interventions in complex interacting systems, and the ethical values so often involved.

In the early years of this century, the better fuel consumption of diesel engines, compared to gasoline engines of the same power, led to a boom in diesel cars and government action to promote them. They are cheaper to run and have lower CO_2 emissions – better for both the purse and the planet. "What's not to like?" However, it was then recognised that diesels emit more of the fine particles and toxic gases that are particularly harmful for the lungs – worse for public health, particularly in cities.

Most discussion of 'climate change' is, like ours above, primarily concerned with the effect of reintroducing long-buried CO_2 to the atmosphere by burning fossil fuels. This is made worse by permanently clearing large areas of long-established forest, which remove significant quantities of CO_2 from the air, perhaps to produce food or fuel. Burning even sustainably produced 'biofuels' releases CO_2 that was only recently absorbed by plants or trees, expecting it will be reabsorbed when those plants are regrown. This is 'good' for climate change, but there are downsides for the environment: heating your house by burning wood produces smoke laden with dangerous particles and toxins – a problem in densely populated areas.

Many types of plant material contain oils or sugars that can be fermented to alcohol, providing good, clean fuel for cars, with no fossil CO_2 emissions – unless, of course, government subsidies on growing those plants encourage farmers to cut down forests or stop growing food crops, raising food prices. The *Bowland Maths* project *You Reckon?* [5G] includes a task which asks students to investigate the pros and cons of this.

Nuclear power has, traditionally, been anathema to the conservation movement. The long-term problem of dealing with spent fuel and decommissioned reactors and the small risk of potentially catastrophic accidents are a grave concern. However, nuclear fission does not release greenhouse gases, and nuclear fuel can generate much more energy than a comparable amount of conventional fuel [5H]. Moreover, burning coal definitely releases CO_2 and other pollutants, including mercury *and* radioactive elements: in fact, coal burning currently releases more radiation than nuclear power stations (McBride et al., 1978) [5J].

A more relevant debate now is to compare nuclear power, which can provide large quantities of continuous power from a single plant but which may take decades to build, with 'renewables' such as wind and solar. Their power production varies with the weather and requires many smaller installations to produce comparable power, but they can be built relatively quickly.

An environmental debate for the mathematics classroom

Focusing on pollution and environmental issues rather than climate change, *Muddying the Waters* [5K] is one of the more adventurous of the Mathematics Assessment Project's *Classroom Challenges*. It is about modelling the pattern of pollution changed by building a dam near a factory (Figure 5.5.1). The lesson includes role-playing discussions between factory owners, the owners of the Riverside Centre, prosecutors, and environmental officers and critiquing the arguments and scientific evidence, quantitative and otherwise, presented in a fictitious court case – an exercise in critical thinking.

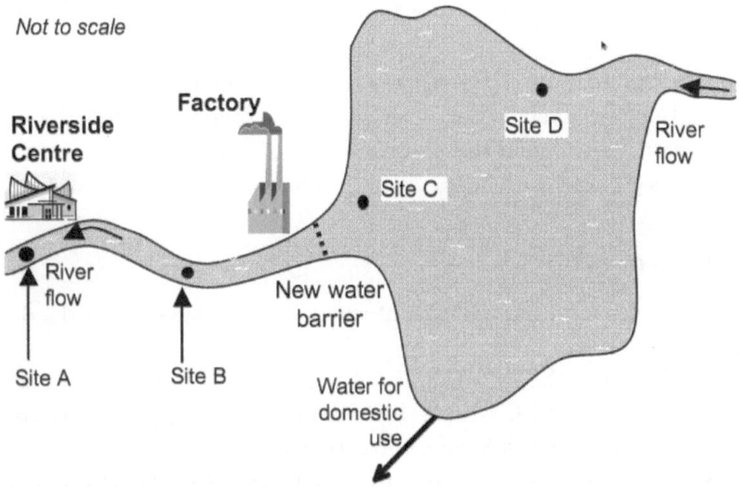

FIGURE 5.5.1 Extract from *Muddying the Waters.*

Source: Mathematics Assessment Project [5K].

5.6 Strategic issues and a critical thinking approach

This book is about how mathematics education can improve mathematical literacy. How can it best contribute in the huge field of the human effects on the environment, with its complex interacting phenomena? Trade-offs, and therefore value-based decisions, are centrally involved. There are mismatches of scale both in space and in time: local short-term needs versus global decade-by-decade changes. Established habits and political systems regularly overmatch cumbersome international structures like the 'Conferences of the Parties' (COPs), which require unanimity.

This strategic issue is a central concern of a community that takes a critical thinking approach to the epistemological and ethical issues involved. Barwell et al. (2022) point out that the human species is privileged over all others in current thinking on the living world; instead, they commend a more symmetrical 'dialogic' approach with ethics based on our answerability and responsibility in relation to all parts of the living world. This perspective stresses that all parts of the ecosystem, including humans, are in a relationship with each other.

For those concerned with mathematical literacy in this context, mathematics education is understood as including the conditions for learners to explore the living world, with suitable guidance, and to become fluent in the appropriate forms of enquiry. Ecosystem crises, including which actions seem needed, become situations to be explored and understood with an emphasis on participation.

This reflects the central message of this book: that mathematics education must go beyond skills and knowledge of mathematics to prepare students to be engaged citizens, able to engage with uncertainty, values-based decision-making, and democratic debate. Hauge and Barwell (2022) propose three principles in support of such an approach: exploring meaningful situations of risk and uncertainty; exploring both scientific/mathematical concepts and societal perspectives; and exploring and learning through dialogue. The need for attention to risk and uncertainty and quantitative assessment of change in this context is clear for the multiple ecosystem crises we face.

Acknowledgements

Our thanks to Richard Barwell for informative exchanges about the dialogic perspective.

Figure 5.5.1 from the Mathematics Assessment Project is available under the Creative Commons CC BY-NC-ND 3.0 licence and appears here courtesy of the Bell Burkhardt Daro Shell Centre Trust.

Thanks to NASA, NOAA, and other agencies for the freely available graphs and data used extensively in this chapter. Figure 5.2.1 is from Levendis et al. (2020) published by the Royal Society under the Creative Commons CC-BY-4.0 licence. All of these institutions offer reputable data on climate change issues – any mistakes or misrepresentations in presenting this data here are fault of the (non-climate scientist) authors of this book.

References

Arrhenius, S. (1896). XXXI. On the influence of carbonic acid in the air upon the temperature of the ground. *The London, Edinburgh, and Dublin Philosophical Magazine and Journal of Science*, *41*(251), 237–276.

Barwell, R., Boylan, M., & Coles, A. (2022). Mathematics education and the living world: A dialogic response to a global crisis. *Journal of Mathematical Behavior*, 68, 101013. https://doi.org/10.1016/j.jmathb.2022.101013

Hauge, K. H., & Barwell, R. (2022). Education for post-normal times. In R. Herheim, T. Werler, & K. H. Hauge (Eds.), *Lived democracy in education: Young citizens' democratic lives in kindergarten, school and higher education* (pp. 65–76). Routledge.

Levendis, Y. A., Kowalski, G., Lu, Y., & Baldassarre, G. (2020). A simple experiment on global warming. *Royal Society Open Science*, 7(9), 192075. https://doi.org/10.1098/rsos.192075 [5B]

MacFarling Meure, C., Etheridge, D., Trudinger, C., Steele, P., Langenfelds, R., van Ommen, T., Smith, A., & Elkins, J. (2006). Law Dome CO_2, CH_4 and N_2O ice core records extended to 2000 years BP. *Geophysical Research Letters*, *33*(14), L14810. https://doi.org/10.1029/2006GL026152. Data set retrieved from https://www.ncei.noaa.gov/access/paleo-search/study/9959

McBride, J. P., Moore, R. E., Witherspoon, J. P., & Blanco, R. E. (1978). Radiological impact of airborne effluents of coal and nuclear plants: Radiation doses from airborne effluents of a coal-fired plant may be greater than those from a nuclear plant. *Science*, *202*(4372), 1045–1050.

Sawyer, J. S. (1972). Man-made carbon dioxide and the "Greenhouse" effect. *Nature*, *239*, 23–25.

Supran, G., Rahmstorf, S. & Oreskes, N. (2023). Assessing ExxonMobil's global warming projections. *Science 379*(6628). https://doi.org/10.1126/science.abk0063. Retrieved from https://www.science.org/doi/10.1126/science.abk0063 [5E]

Thom, R. (1977). Structural stability, catastrophe theory, and applied mathematics. *SIAM Review*, *19*(2), 189–201.

Links to useful material

To visit any of these links, scan this QR code or visit ltml.mathlit.org – append the link code to go directly to the entry – for example, ltml.mathlit.org/5A

The original source links are given below for attribution purposes:

[5A] *Global Climate Change* – NASA
https://climate.nasa.gov/vital-signs/global-temperature/

[5B] *A Simple Experiment on Global Warming* – Royal Society Open Science
https://royalsocietypublishing.org/doi/10.1098/rsos.192075

[5C] *Showing that Carbon Dioxide Is Made When a Hydrocarbon Is Burned*
https://youtu.be/6l8D6DsRDYk
https://youtu.be/6o8QXvKZ2xI
https://youtu.be/b7e9QjNX51s

[5D] *Global Warming Potential* – Wikipedia
https://en.wikipedia.org/wiki/Global_warming_potential

[5E] *Assessing ExxonMobil's Global Warming Projections* – Science
https://www.science.org/doi/10.1126/science.abk0063

[5F] *Trends in Atmospheric Carbon Dioxide* – NOAA Global Monitoring Laboratory
https://gml.noaa.gov/ccgg/trends/

[5G] *You Reckon? Bio Fool* – Bowland Maths
https://www.bowlandmaths.org/materials/projects/online/you_reckon/You%20Reckon_Web/page_17.htm

[5H] *Energy Density of Uranium* – various sources
https://www.plux.co.uk/energy-density-of-uranium/
https://atomicinsights.com/energy-density-comparison/
https://www.euronuclear.org/glossary/fuel-comparison/

[5J] *Do Coal-fired Power Stations Produce Radioactive Waste?* – Scientific American
https://www.scientificamerican.com/article/coal-ash-is-more-radioactive-than-nuclear-waste/

[5K] *Muddying the Waters* – Mathematics Assessment Project
https://www.map.mathshell.org/lessons.php?unit=9400&collection=8

6

PLANNING FOR GOOD THINGS IN LIFE

We all know, in principle, that careful planning of what we do with our time can enable us to get more out of life. But the planning itself takes time – and for many people it is an uncongenial activity. They would rather "be getting on with life" without having to spend time planning the details of what to do when, and how to make it happen. Yet some things obviously need to be planned – or else frustration, or worse, is sure to follow. So what kinds of planning are right for you?

ACTIVITY

Ask small groups to discuss what situations they plan for – and for what other situations they think it might be worth it.

Lifestyle choices and opportunities are obviously determining factors in the range of options considered – the more choices you have, the more decisions you have to make as to what to do, and when. Some total 'spur of the moment' people don't plan, but if you have some time for 'slow thinking', then considering what and how to plan is worthwhile.

All work environments involve planning of various kinds. For junior employees their work may be largely imposed 'from above', though in any well-run company there is space for all employees to use their intelligence in making the management's plan work as well as possible. Promotion leads to increased responsibility for planning decisions – for example, in detailed

DOI: 10.4324/9781003303503-7

scheduling and in assigning people to tasks in a way that reflects their different capabilities.

In this chapter we first look in Section 6.1 at the challenges in organising events and how mathematics can be useful. Section 6.2 outlines some well-engineered teaching materials focused on planning. Section 6.3 looks at mathematics to tackle the more general problem of optimising a situation within constraints, extending this in Section 6.4 to the science and mathematics of logistics – with a forward look to data science, to which we return in Chapter 7. The chapter includes a mix of 'know how' and 'know about' mathematical literacy mixed with critical thinking. Planning is something nearly everyone must know how to do, and it is also useful to know about the sophisticated tools available for complex situations.

6.1 Organising events

The *Airplane Turn-round* and *Table Tennis Tournament* tasks [6A] introduced in Chapters 1 and 2 are typical of a genre: organising events or, more widely, processes. Other familiar examples are as diverse as getting the family up and ready for school or work, or organising a school event, an election, or a strike. The ability to organise is a valuable high-level skill. What are its main elements?

- The participants
- The physical environment in which the event will take place
- The sequence of activities and the rules that govern them

The challenge is to devise a systematic approach which addresses all the significant variables and, one hopes, all eventualities that might arise. Usually, the organiser will have knowledge of similar events – even experience in taking part – but they are unlikely to have thought through all the aspects of planning required.

It is easy, and usually disastrous, to forget to carefully consider the interests and feelings of the people who will be involved; the study of 'human factors' is a field of its own, beyond the scope of this book except to underline the importance of trying to get everyone 'onside'. In the same way, a congenial physical environment is an important factor in the success of any event. These matters are not mathematical beyond being included in a systematic approach.

For the sequence of activities and the rules that govern them, simple mathematics is helpful. Things like lists and tables are essential. Less familiar types of diagrams can clarify the situation. For example, the *Airplane Turn-round* task (introduced in Section 1.3) lists seven operations taking a total of 160 minutes (Table 6.1.1) and asks how this time can be minimised. In this

TABLE 6.1.1 Operations needed to 'turn-round' a passenger aircraft

Operation	*Time (minutes)*
A. Disembark passengers	10
B. Clean the cabin	20
C. Refuel the plane	40
D. Unload the baggage	25
E. Embark new passengers	25
F. Load new baggage	35
G. Final safety check	5

Source: Adapted from Bowland Maths [6A].

case, with the realisation that some of the jobs can be performed in parallel, one can sort out the minimum time needed by some thought and some trial and error.

ACTIVITY

Ask small groups to tackle the Airplane Turn-round task individually then discuss how they did it.

Systematic representations – Pert diagrams and Gantt charts

There are various systematic methods for tackling scheduling problems of this kind. The 'Pert diagram' in Figure 6.1.2a shows one systematic way to record which steps must be sequential and which can be done in parallel, and the times involved. There is a formal Program Evaluation and Review Technique (Pert) for using such diagrams in more complex situations.

The total turn-round time is determined by the path through the diagram with the longest total time – in this case 65 minutes for the steps of unloading and loading baggage and the final check. So, efforts to shorten turn-round time should focus on the steps on this *critical path*. Knowing to identify the critical path, and try to shorten it, is a powerful general insight – and strategy.

If, for safety reasons, it was decided that refuelling should not take place while any passengers are on board (as with rockets) the modified diagram Figure 6.1.2b would apply. The critical path changes so the turn-round time becomes 80 minutes – an additional operating cost for the airline.

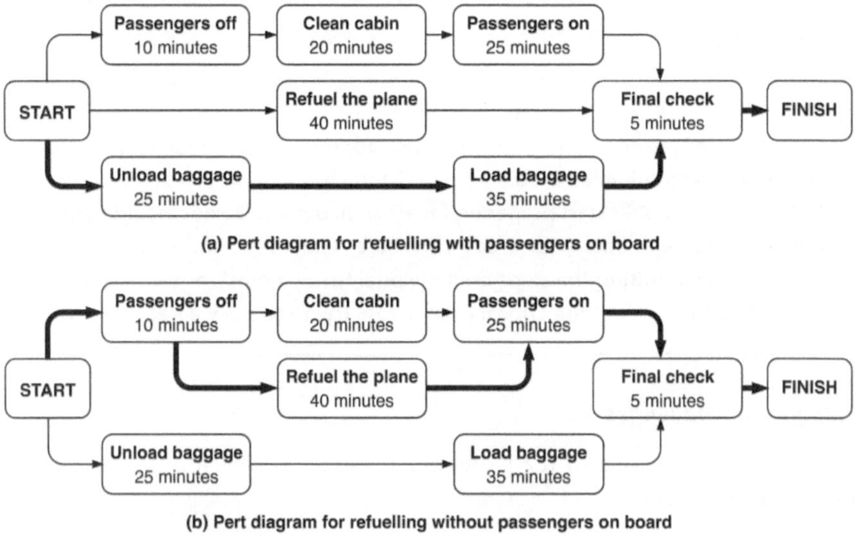

FIGURE 6.1.2 Simplified Pert diagrams for refuelling (a) with and (b) without passengers. 'Critical paths' shown in bold.

Cooking and serving a large meal with many dishes is a situation of similar complexity that many people will have grappled with.

Another planning tool in regular use is a Gantt chart. Gantt charts arrange the tasks along a timeline, which makes them useful in organising the details of complex projects and monitoring their progress. Figure 6.1.3 provides a simple example, based on Figure 6.1.2b. Gantt charts can be constructed in spreadsheets, though specific products with extra functionality are available. These examples again show the value of customised mathematical representations in giving power over complex situations.

	Duration	Start	End
Unload Baggage	25	0	25
Load baggage	25	25	50
Passengers off	10	0	10
Clean cabin	20	10	30
Refuel the plane	40	10	50
Passengers On	25	50	75
Final check	5	75	80

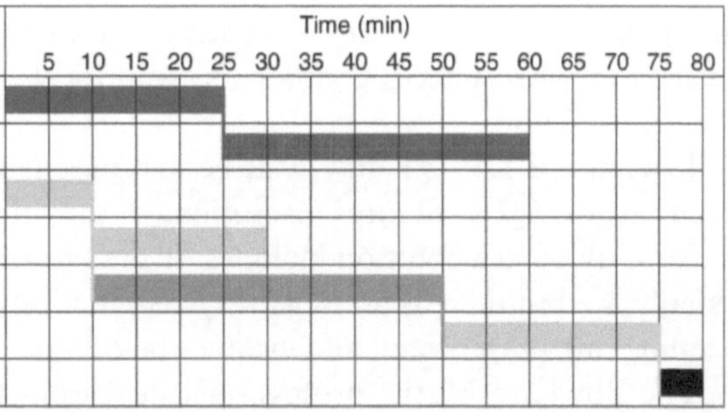

FIGURE 6.1.3 Airplane turn-round – a Gantt chart.

ACTIVITY

Plan how to cook and serve a meal on time.

Choose recipes for a three-course meal (appetiser, main course, and dessert) and draw a suitable Pert diagram or Gantt chart. Identify the critical path and when you should start preparing for a 7 p.m. dinner. Do not include preparation for components (e.g. a cake) that are ready in advance. It will save time to start by writing the steps on individual 'sticky notes' (or the computer equivalent) to work out the critical path before trying to draw a neat diagram.

Pedagogical challenges

The following are some of the issues teachers face in developing these skills for organising events:

- How open should the task be?
- How much time should be devoted to the activity?
- How should support be given to those who struggle with the task, or when their work no longer seems productive?
- What is the payoff in learning?

These are much the same challenges as arise when solving non-routine problems within mathematics. Let us look at them in the context of organising a tournament. Teachers may feel that the version of *Table Tennis Tournament* in Chapter 2 (Figure 2.1.2) – which simply asks students to plan a tournament – is too open, though it will often arise in this vague form in real-world situations, leaving the organiser to make it more specific. A lightly scaffolded version designed for use in a single lesson is shown in Figure 6.1.4.

Notice the questions that an organiser should ask in embarking on this task are not only raised but answered in this version. This reduces the strategic demand of the task, assists discussion by setting common parameters, and gives the teacher a more predictable lesson, and certainly makes for an easier job when checking students' solutions! The version in Figure 6.1.5 takes this further.

In Figure 6.1.5, Q1 may well be just an exercise, testing skills that have been taught. Q2 structures a solution, setting out the parameters and showing exactly what a solution includes. It can be tackled by trial and error – and usually is – but there is an elegant systematic, generalisable solution based on keeping one player fixed and cycling the others.

The Bowland Maths professional development module on *Tackling unstructured problems in the classroom* [6A] begins with an introductory session in which teachers discuss the issues, compare more or less structured

You have the job of organising a table tennis league.

- 7 players will take part.
- All matches are singles.
- Every player has to play each of the other players once.
- There are four tables at the club.
- Games will take up to half an hour.
- The first match will start at 1.00 p.m.

Plan how to organise the league, so that the tournament will take the shortest possible time.

Put all the information on a poster so that the players can easily understand what to do.

FIGURE 6.1.4 Table tennis tournament – a lightly scaffolded version.

Source: From Bowland Maths: *Tackling Unstructured Problems* [6A].

You have the job of organising a table tennis tournament.

- 7 players will take part
- All matches are singles.
- Every player has to play each of the other players once.

1. Call the players A, B, C, D, E, F, G
   Complete the list below to show all the matches that need to be played.

   A v B   B v C   ....
   A v C   B v D   ....
   ....

2. There are four tables at the club and each game takes half an hour.
   The first match will start at 1.00pm.

   Copy and complete the poster below to show the order of play, so that the tournament takes the shortest possible time.
   Remember that a player cannot be in two places at once!
   You may not need to use every row and column in the table!

Start Time	Table on which the game will be played			
	1	2	3	4
1.00	A v B			
1.30				
2.00				
2.30				
3.00				
3.30				
4.00				
4.30				

FIGURE 6.1.5 Table tennis tournament – a directive version.

Source: From Bowland Maths: *Tackling Unstructured Problems* [6A].

versions of the task, watch a lesson video, and then together prepare a lesson based on a relatively open version of the task to teach. After observing how the activity works, they reflect in a structured follow-up session on what happened and what they have learned about handling non-routine problems in the classroom.

6.2 A case study for teaching planning and organising

This section gives an example of materials to develop planning skills. The *Numeracy through Problem Solving* project (NTPS) was described in Section 2.4. Three of the five modules, *Design a Board Game*, *Produce a Quiz Show* and *Plan a Trip*, target planning skills. For those who want a well-supported entry into teaching for mathematical literacy, these modules, and their four-stage design, may be useful.

Each module works on a group-project basis and has four stages:

- Stage 1: *Explore the domain* by working on and evaluating exemplars provided.
- Stage 2: *Generating and sifting ideas*, then reviewing them and deciding what to take forward.
- Stage 3: *Developing a 'product'* in detail.
- Stage 4: *Evaluating* what other groups have produced – then reviewing your own.

Produce a quiz show

Many websites provide support for game show simulations of various kinds. In the *NTPS* module, student groups devise, schedule, run, and evaluate their own game shows. This involves: preparing, timing, and testing questions (using number and statistical concepts); planning room layouts; and scoring systems. There are four stages:

- Stage 1: Groups of students take it in turns to act out a number of TV-type quizzes that are provided, identifying and commenting on faults and shortcomings in the organisation, rules, questions, scoring systems, and presentation. These games are designed with faults – some obvious, others less so; using established game shows as a model would set a standard that can be discouraging.
- Stage 2: In groups, students share ideas for their own quiz, reach agreement on which to develop, and draw up a plan of action.
- Stage 3: Each group prepares, tests, and organises its questions, scoring systems, rules, and final running order. Groups decide how the furniture and equipment will be arranged during the show.

- Stage 4: Groups take it in turns to present their quizzes, with the rest of the class providing competitors – and the audience. Afterwards, each quiz is evaluated, first by other members of the class, and then by the group who produced it. A further opportunity may be given for a group to enact their quiz with different groups of contestants – perhaps a different class.

The student booklets are designed to maximise student agency, providing suggestions on essentials only *after* the point where students should have thought of it for themselves. All the *Numeracy through Problem Solving* materials can be downloaded free from the Shell Centre website [6B].

6.3 Optimising within constraints

In planning, there are always constraints of time, money, personnel, and available resources – just to name the most obvious. Often the constraints sharply limit the range of options available, so seeing how far the constraints can be stretched is a strategy for getting a better outcome.

Vacation planning is a common example, where most people have constraints, including those on both time and money. Your job probably comes with a fixed number of paid vacation days off work each year; if you can take your vacation in weeks that contain a public holiday, you can be away from home for longer. Equally starting out on a Friday evening and returning just in time for work on a Monday morning will extend your vacation. But paying for the extra accommodation may challenge your money constraint.

In this way, different constraints compete for priority, which leads to the question of what you are seeking to optimise. Suppose you are looking for reliable sunshine, warmth, and a beach. That will eliminate a lot of nice places and help to rank order those that satisfy this constraint. When you consider the cost, and also want to be in an interesting place, other trade-offs come in. How are these things ('goods' to economists) to be balanced?

Most of us who try to be somewhat systematic would, when planning such a vacation,

- explore the range of possibilities, for example by browsing the web and talking to friends;
- choose some 'candidates' that fit within our 'tight' constraints;
- gather information on cost, weather, and attractions; and then
- make a choice based on 'an overall impression'.

This method seems to satisfy people – the main aim. Discontent usually arises when the data are wrong: the flight is delayed, the hotel is still under construction, or the "nearby beach" on the website is a mile away.

Systematic approaches

For situations where more formal, well-defined, methods of choosing are needed, there are various approaches based on two fundamental concepts: *boundaries* and *utility*. The concept of boundaries is a relatively straightforward one – tight constraints that must be satisfied by any solution. The PISA item *Which car?* in Figure 2.2.1 is an example where the boundaries alone determine the choice. More often there are many possible choices that fit the boundaries; then the concept of a *utility function* comes in. This is a function of all the significant variables that reflect the values of the person or other entity making the decision. For some situations, like those focused on money which we will address in Chapter 9, the utility function may be relatively straightforward. The directors of a British or American public company are currently required by law to focus predominantly on 'shareholder value' – so the share price is the utility function. Some jurisdictions also include the interests of other 'stakeholders', both employees and society at large, so deciding on the balance between competing 'goods' is part of the management challenge.

In such situations, a function must be built from the significant variables. Unsurprisingly, a linear function is often preferred, so the utility U is defined by

$$U = w_1v_1 + w_2v_2 + w_3v_3 + \cdots + w_nv_n$$

where w_i is the (numerical) *weight* assigned to the variable v_i.

ACTIVITY

Ask small groups to apply boundaries and utility to planning a short vacation:

- Set three boundary constraints for the vacation (e.g. maximum cost per person, dates, main activity) and choose at least three utility variables (e.g. travel time, daily temperature, live music venues).
- Find five alternative vacations that fit the boundary constraints (alternatives might be variations, e.g. in dates).
- Compare the alternative vacations using the informal approach described above and choose one.
- Construct a set of weights for each variable that will make the preferred choice have the highest utility. (*Hint:* In deciding weights it is the *variation* in each variable that determines its overall influence on U, not its absolute value).
- Discuss the experience.

> Phil and Cath plan to make and sell boomerangs in two sizes: small and large.
>
> Phil will carve them from wood, taking 2 hours for a small one and 3 for a large one.
> Phil has a total of 24 hours available for carving.
>
> Cath will decorate them.
> She only has time to decorate 10 boomerangs of either size.
>
> The small boomerang will make \$8 profit, the large boomerang \$10.
>
> They want to make as much money as they can.
> **How many small and large boomerangs should they make?**
> **How much money will they then make?**

FIGURE 6.3.1 Maximising profits: selling boomerangs.

Source: Mathematics Assessment Project [6C].

This domain of mathematics, *linear programming*, is a standard topic in some high school curricula. Perhaps surprisingly, the core result is that the boundaries usually play a central role in determining the optimum.

Boomerangs – an optimisation activity

The lesson *Maximising profits: Selling boomerangs* illustrates this kind of planning. The essence of the problem is in Figure 6.3.1. The goal is maximising profits (the utility function) within two boundary constraints of time available.

The materials on the MAP website [6C] provide detailed support, with examples of student work for discussion in class. It assumes that students have not yet been taught linear programming, so this is a non-routine problem for them to solve. Most students tackle this novel problem by enumerating possibilities and then organising them in a table – a more challenging task than one might think. The four examples of student work in the module, and the commentary on them, are illuminating.

A more sophisticated approach using algebra, with x small and y large boomerangs, expresses the profit and constraints as

- $P = 8x + 10y$ (profit)
- $x + y \leq 10$ (Cath's time constraint)
- $2x + 3y \leq 24$ (Phil's time constraint)
- $x \geq 0$ and $y \geq 0$ (no negative boomerangs!)

When there are only two variables, the standard linear programming approach is to graph these linear functions, looking for the maximum value of P which stays with the constraints. Figure 6.3.2 shows how this works – any values of x and y falling in the white area break one or both of the time constraints. The 'permitted' area is the darkest shaded region $ABCD$. $P = 88$

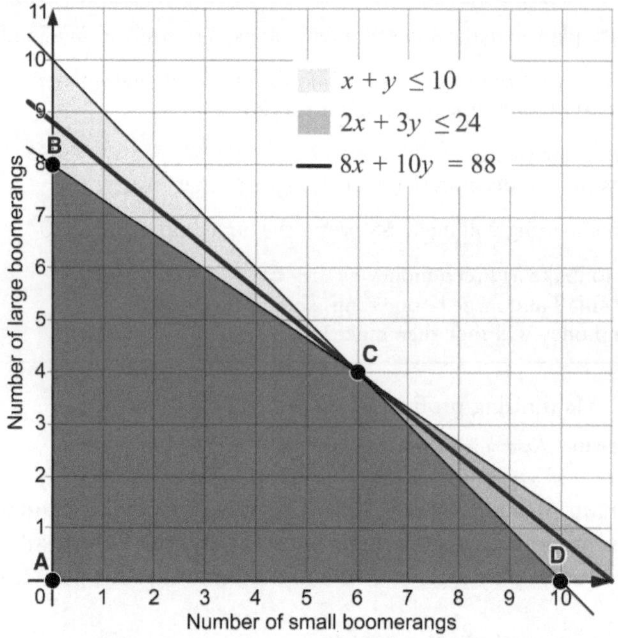

FIGURE 6.3.2 Maximising profit on 'boomerangs' – a linear programming solution.

is the maximum value for which the 'profit' line $P = 8x + 10y$ passes through $ABCD$, just 'kissing' it at point C where there are six small and four large boomerangs.

A graph like that in Figure 6.3.2 can be easily made with free online tools such as Desmos Graphing Calculator or GeoGebra [6D]. The website links to one. First plot the two inequalities, which can be typed in as '$x + y \leq 10$' and '$2x + 3y \leq 24$'. Then, add the profit formula as '$8x + 10y = P$'. Desmos and GeoGebra will automatically create a 'slider' that lets you interactively adjust P, although you will need to specify the slider settings to allow a sufficiently large range for P.

As an extension of this problem, it is interesting to vary the parameters (10, 24, 8, and 10) to see when the optimum moves to a boundary – either all small or all large boomerangs, so at a different corner of the outlined shape – or when there is more than one optimum value of x and y. It is easy to explore these questions using the graphing calculator sliders.

6.4 Logistics and data science

Linear programming is just one approach in a broad field of mathematical approaches to doing things more efficiently. This work goes back to the eighteenth century, but it really escalated during World War II under the name of

Operational Research (OR, also called Operations Research). In 1947 under the auspices of the British Association for the Advancement of Science, a symposium was organised in Dundee. In his opening address, Sir Robert Watson-Watt offered a definition of the aims of OR (Zuckerman, 1964):

> [T]o examine quantitatively whether the user organization is getting from the operation of its equipment the best attainable contribution to its overall objective.
>
> *(p. 288)*

This down-to-earth definition reflected his role, as one of the inventors of radar, in getting as much useful information as possible out of the limited numbers and capabilities of early "radiolocation" sets. The most famous case was in the Battle of Britain where radar played a major role in the efficient deployment of defensive fighter aircraft, which in turn prevented an air invasion of Britain. If you generalise 'equipment' to 'resources', this remains a fair definition of the field now usually called logistics. A great variety of approaches have been developed since then. They involve collecting data, then mixing modelling and data analysis in various ways.

Making sense of data has a long history, often carried out by scientists, statisticians, librarians, computer scientists, and many other professions, as well as mathematically literate citizens in their everyday lives. To take one foundational example, in 1854 there was a cholera epidemic in London. John Snow (1854) collected data and made a 'map' of the address in London where each case occurred – part of his map is shown in Figure 6.4.1. Each bar on the map represents one death at that address. The cluster near the Broad Street water pump alerted him to the cause of the cholera outbreak – the

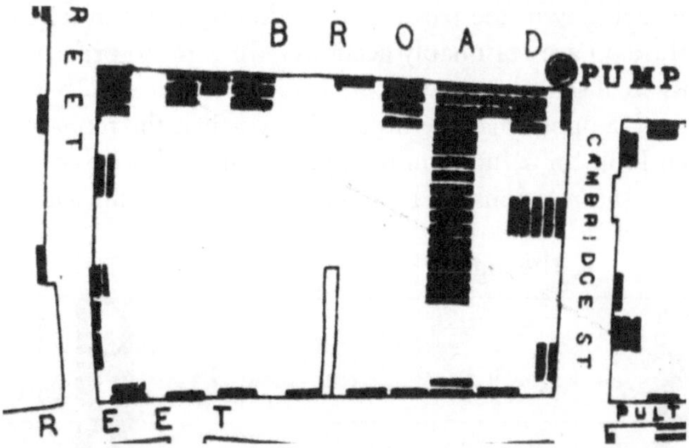

FIGURE 6.4.1 Part of John Snow's map of the 1854 London cholera outbreak. Each bar represents one death at that address [6E].

water supply, rather than "unhealthy air from the river" which was previously thought to be the cause.

This work was the beginning of systematic epidemiology. Snow's vivid display informed how we visualise and analyse data to this day. In the twenty-first century, advances in computer power, the digitisation of the world through the internet, and associated software design (including artificial intelligence) enable data scientists to extract highly complex patterns from enormous quantities of data and so develop probabilistic predictive models that allow better business decisions. But the underlying principles are the same as those applied by John Snow in 1854.

The principles of data science

We illustrate the principles of data science with a very simplified example. Let's suppose we run a small business – an ice cream shop. We will use a data-driven logistics procedure, also called predictive modelling.

Step 1 is to identify your system of interest – a system being just some 'bit' of the world, in our case the ice cream shop with its surroundings, and its possible customer base of people who live near or travel by. We also need to identify which variables – also called properties, or attributes – are important for our system. For the ice cream shop, the number of ice creams sold would be of central importance. Other variables, such as the weather, are also likely to be important.

Step 2 is to measure the system, to collect data. This could be as simple as recording the number of ice cream sales each day, along with the daily temperature, including the forecast for Friday. This data is stored in a spreadsheet on a computer, or in a database, as in Figure 6.4.2a.

Step 3 is to find a mathematical function which takes in the things we know and predicts things we want to know. We know the ice cream sales and temperature each day in the past – and we know temperature forecasts for tomorrow (Friday) are reasonably accurate. We want to predict the number of ice creams we'll sell tomorrow as indicated in Figure 6.4.2b. From experience, we might suppose that the hotter the weather, the more ice cream we sell. We then look for a function (or model – here they're the same thing) that expresses the relationship between temperature and ice cream sales.

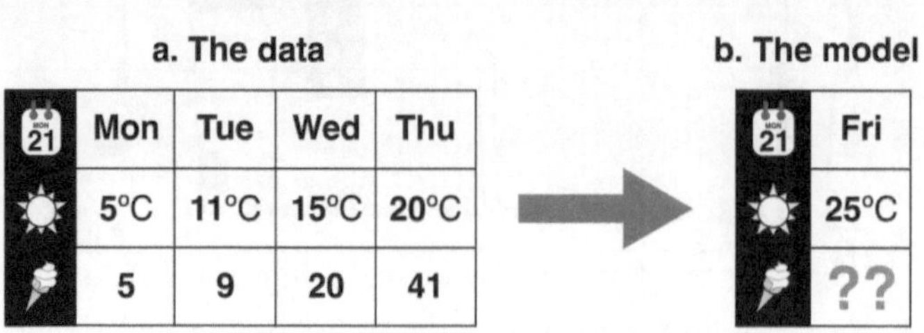

FIGURE 6.4.2 The essentials of data science.

Temperature (°C)	Sales Data	Sales Model
5	5	
10	9	
15	20	
20	41	41
25		82
30		164

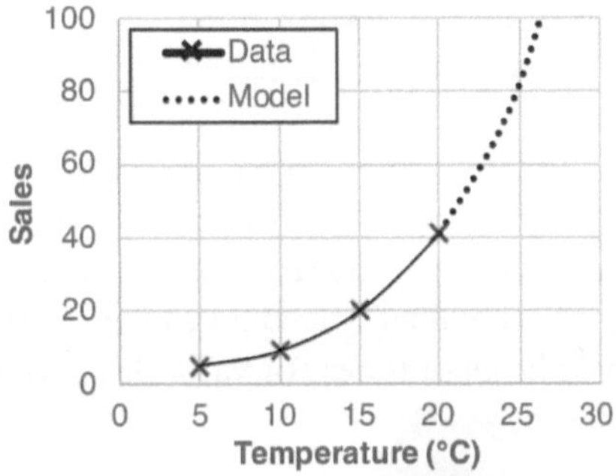

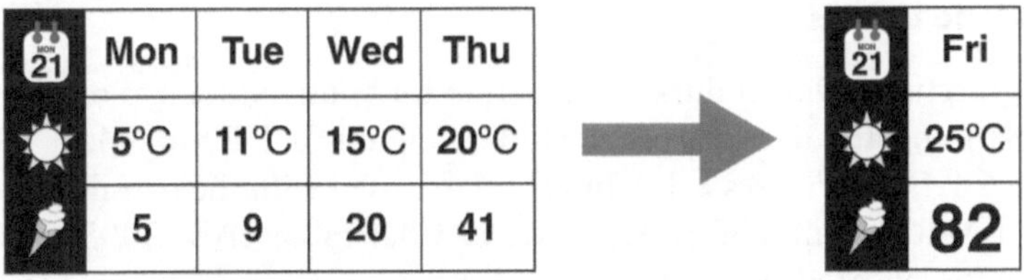

FIGURE 6.4.3 A predictive model for ice cream sales.

We are free to find this mathematical function by any means – experiment with functions using pen and paper or a spreadsheet, a statistician might make statistical models, and a data scientist might use machine learning.

From these four data points, it is clearly not a linear function – indeed, it looks as though the sales roughly double for every 5°C increase which would predict 82 sales of ice cream on Friday, shown numerically and graphically in Figure 6.4.3.

From a data science perspective, how we find this function is not fundamentally important; what matters is whether the function makes sufficiently accurate predictions. We can test a candidate predictive function by using it to "predict" our historical data, and then see how closely its predictions match our records of the past. As in all data-driven research, it is important to generate the hypothetical model with one data set and then test it on another one. In this case, we might test the model on data from a friend's ice cream shop. This key principle is often ignored. (Of course, this four-point data set is ridiculously small and only used to illustrate the principles.)

Step 4 operates when we find a predictive function that predicts with sufficient accuracy. Then the final step of the process is to use this model to predict the future and make better informed decisions. In our case, we could predict how many ice creams we will sell on Friday. This could be of great value to our decision-making; we could check to see if we have enough ice cream in stock or need to order more, we might roster more staff to work today, or perhaps on other days not bother opening at all because we don't expect enough sales!

A data scientist might extend this model to include many more variables – the location of the ice cream store, nearby events such as a local park fair, which flavours sell best, or online marketing activity, with the hope that this would give a function with better predictive power. The difference between humans making estimates in their head, researchers in Operational Research, scientists forming theories of the world, and machine learning practitioners is only a difference in the tools they use and the scale of the enterprise. The objectives and procedural frameworks are the same. Modern data science, appropriately called 'big data', enables these principles to be extended to extremely large numbers of variables and data points.

Acknowledgements

We are grateful to Daniel Burkhardt Cerigo for contributions to our thinking about planning and data science, including Figures 6.4.2 and 6.4.3.

Table 6.1.1 and Figures 6.1.4 and 6.1.5 are used with the permission of the Bowland Maths maintainers. © Bowland Charitable Trust 2008. Bowland Maths materials are free for educational use.

Figure 6.3.1 and extracts from *Selling Boomerangs* from the Math Assessment Project *Classroom Challenges* [6C], courtesy of Bell Burkhardt Daro Shell Centre Trust. The full materials are available under the Creative Commons Attribution, Non-commercial, No Derivatives Licence 3.0.

Figure 6.4.1 is taken from a version of Snow's (1854) map digitised by the UCLA Department of Epidemiology – see link [6E] for the full version.

References

Snow, J. (1854). *On the mode of communication of cholera*. C.F. Cheffins, Lith.

Zuckerman, S. (1964). In the beginning – and later. *OR*, *15*(4), 287–292. https://doi.org/10.2307/3007115.

Links to useful material

To visit any of these links, scan this QR code or visit ltml.mathlit.org – append the link code to go directly to the entry – for example, ltml.mathlit.org/6A

The original source links are given below for attribution purposes:

[6A] *Tasks from Professional Development Modules* – Bowland Maths
Table Tennis Tournament:
https://www.bowlandmaths.org/materials/pd/online/pd_01/pd_01_class.html
Aircraft Turn-round: https://www.bowlandmaths.org.uk/materials/pd/online/pd_05/pd_05_class.html

[6B] *Produce a Quiz Show* – Shell Centre for Mathematical Education
https://www.mathshell.com/materials.php?&series=numeracy
[6C] *Maximising Profits: Selling Boomerangs* – Mathematics Assessment Project
https://www.map.mathshell.org/lessons.php?unit=9205&collection=8
[6D] *Graphing Tools for Linear Programming* – Desmos & GeoGebra
https://www.desmos.com/calculator
https://www.geogebra.org/calculator
[6E] *Map of Cholera Cases in London* – Snow (1854)
https://www.ph.ucla.edu/epi/snow/highressnowmap.html
https://https://en.wikipedia.org/wiki/John_Snow

7

LOOKING PAST THE 'SPIN'

Every day we are bombarded with messages from people who would like to sell us something – everything from products we don't need but might enjoy to political ideas that might persuade us to support their party or their cause. They often set out to make their case more persuasive by quoting numbers or displaying graphs that support it. But these are, inevitably, carefully selected from a very much larger set of data. How does a mathematically literate person best think through the options implied, to 'buy' or not to buy – as ever, without wasting time? That is the theme of this chapter. Much of what follows will be familiar to the reader, but we hope a broad overview will be thought-provoking.

Context is always an important part of mathematical literacy, but particularly so here – where mathematics is used to add credibility to an argument. 'Looking past the spin' involves *critical thinking* that *takes the context seriously* – looking in depth at the context, the language used to present the argument, and how mathematics is being used to support it. Is the mathematics correct in itself? Are the assumptions behind the mathematics, such as the data chosen, valid? Does the language used fairly describe what the mathematics shows? Consequently, this chapter spends significant time discussing the wider background of topics where debate is frequently supported by the use, or abuse, of mathematical data.

With an exception that we will return to later, the people who design marketing – whether in business or politics – usually avoid 'the lie direct'. To present as a fact something that can be shown to be untrue runs two risks: rebuttal by a trusted 'fact checker' and, much more seriously, damage to your reputation as a reliable source. That is a long-term injury that few organisations think worth risking for a short-term gain in sales or influence.

DOI: 10.4324/9781003303503-8

7.1 Comparing products

Let us start with the most obvious form of marketing. Advertisements for 'consumer products' assail us on the street, in every corner of the internet, and of our television and screen viewing. Even non-commercial stations like the BBC advertise their own programmes extensively.

From the marketers' point of view, this is a public service, letting us know about things we might well have missed. This is undoubtedly true and potentially valuable. But from the public's point of view, it is not something we can avoid or easily control. It sometimes feels, in a memorable phrase, 'like cocktails from a firehose'. Friends add to this with accounts of things they have bought or looked at. How does the mathematically literate person handle this stream of information – and disinformation?

We examine in Chapter 9 the issue of 'needs' versus 'wants'. Here we just assume that you are well aware of your needs. If an advertisement should remind you of a need that you have forgotten 'to put on the list', so much the better – a real public service, epitomised in 'public service announcements' designed to warn of risks or encourage healthier living. But, beyond needs, what about things you would simply like to own or do?

Most people have a 'small' sum of money that they are prepared to spend without worrying about it. For things under that limit, it is time-effective just to buy it. Of course, you will review your limit from time to time – particularly if your circumstances change. You will also have to consider how 'small' outlays accumulate over time – there are marketing practices that exploit this by advertising a modest-sounding subscription, or payment in instalments.

If you have created a weekly or monthly budget for 'discretionary spending' – on clothes, tickets, or whatever – some agonising may be in order. Indeed, it may be part of the enjoyment, perhaps with friends, to talk through various things you might buy. But again, many would feel careful consideration of alternatives would spoil the fun. Personal budgeting of this kind is discussed in more detail in Chapter 9.

For more substantial purchases, finding the right product at the right price demands more thought. The potential saving from 'shopping around' for a new smartphone or television could fund another substantial 'want'. But looking at alternatives, evaluating their quality and comparing prices is potentially a huge task. Fortunately, there is help at hand. The traditional sources are consumer magazines and their websites – sites such as *Consumer Reports* in the United States, *Which?* in the United Kingdom, *Choice* in Australia; these have established programmes of regular research-based comparative reviews of consumer items from coffee grinders to cars – but, typically, you will need a paid subscription to get beyond the free overview. These have been joined by a large number of product comparison websites offering professionally written reviews – although, since these are typically funded by advertisements, possibly for the very products they are reviewing, users need to assess their impartiality.

This 'traditional' approach to product reviewing now faces its own competition. With the increase in online shopping, 'crowd-sourced' user reviews and ratings have become ubiquitous – and being able to hear the honest views of genuine people who have actually bought and used the product sounds wonderful – but some are fake reviews from 'users' who are given incentives to write positive reviews, or negatively review a competitor (see, for example, The Secret World of Fake Online Reviews, *The Guardian*, 22 April 2023 [7A]). Even without deliberate fakery, there are sources of bias – a proportion of all goods are faulty and someone angry at receiving a faulty unit may be more inclined to write a review than many users who thought the product was OK. Others may rush out a glowing review based on first impressions that misses a serious flaw in the product. "Star" ratings are popular – but what do they mean? Some people regard anything less than five stars as a failure, others will see anything more than three stars as acceptable and reserve five stars for exceptional products – and if a product has hundreds of reviews, it is tempting to go by the average rating. Table 7.1.1 parodies the problem of superficial five-star reviews.

More recently, social media have become a leading source of product information for consumers – and, hence, a vital marketing tool for business and politics. Initially the domain of enthusiastic amateurs, being a social media 'influencer' can now be a lucrative full-time business with celebrity status, utilising professional videos with high production values. A good influencer will be totally transparent about sponsorship and how they obtained the products they feature. They often produce in-depth reviews rivalling the traditional consumer advice publications. Others will give the impression that these are products that they use personally (true or false) which they are recommending to you as a friend.

TABLE 7.1.1 How do you summarise reviews?

Average customer review:		★★★★½
Nev	Loved the colour	★★★★★
Kat87	Arrived next day, well packaged	★★★★★
Wurzel	Fits like a glove, came with free pen	★★★★★
Joolz	Does what it says on the tin	★★★★★
Fluff77	Looks super, but batteries not included in box	★★★★
JWSmith	Exploded – second-degree burns. House burned down. Posting this from hospital	★

For illustrative purposes only – as they say in the ads.

The internet has brought another, often neglected, option, especially when buying 'tech' products or appliances: it is now often possible to download the instruction manual, which often gives clearer, more detailed information about the product's specifications and capabilities than the marketing material.

How can you make objective decisions?

The wide range of products and the variety of features that are so often on offer can make thinking through alternatives in a systematic way a challenge. The mathematical challenge of integrating the quality assessments from reliable sources with the prices on offer is modest – basically one of organising information. 'Irresistible offers' at the 'sales' complicate the process only slightly.

Whereas systematic reviews are objectively worthwhile, it is not surprising that many of us just look for a good discount from a source we know and have found reliable, and buy.

ACTIVITY

How do you make buying decisions?

Students compare notes on how they make buying decisions at various cost levels.

In Chapter 6 we looked at a systematic method of comparing things with various desirable attributes by adding individual scores (v_n) with weights (w_n) chosen to reflect the importance of that factor – and how much those scores vary – defining a 'utility' function:

$$U = w_1v_1 + w_2v_2 + w_3v_3 + \ldots + w_nv_n$$

The same approach is used in evaluating products. Looking back at the simplistic averaging in Table 7.1.1, you might feel that product safety deserves more weight than colour or packaging. To take a real example, an article from *Choice* [7B] describes how they rank washing machines. For each machine, scores are allocated to performance on five variables: dirt removal (d), rinse performance (r), gentleness (g), water efficiency (w), and spin efficiency (s). Each of these variables would have a separate scoring system. Then machines are compared using the utility function:

$$U = 0.4d + 0.2r + 0.15g + 0.15w + 0.1s$$

The system to decide the winners of the decathlon in track and field athletics is a complex real-world example of a utility function – see this article from NRICH [7C]. There are ten events. Each athlete receives a score for performance (either time or distance) on each event. The scoring systems, defined and revised separately for each event, were originally set up so that a world record performance would receive 1 000 points. For the decathlon, events are equally weighted, so the scores are added to determine the winner. Other sports, like fantasy or real football, offer many examples.

ACTIVITY

Creating and using a utility function.

a How would you define the GOAT (greatest of all time) for the sport, or the music style, of your choice.
First, in groups choose a sport, music style, or some another shared interest. Decide on the contenders and the variables (factors) which make 'greatness'. Decide how each factor should be scored and assign weights to each factor. Then create a U-value for each individual. Discuss whether you agree with the rank order your U-function gives. If not, try changing the weights so that it better represents your values.

b Alternatively, undertake the same activity for products. Choose a 'product category' (e.g. coffee machines, electric luxury cars) for which there are several factors which significantly affect desirability, and on which there is easily accessible data for multiple individual products. For example, for electric luxury cars, factors will include range, style, internal fittings, and price. Proceed as above.

In some countries test scores are used in this way to evaluate teachers and schools. A narrow selection of variables leads to a distortion of what is taught to focus on the content being tested, which only rarely includes all the learning goals in a balanced way – 'what you test is what you get'.

What do marketers do?

It is informative to think about marketing from 'the other side'. The essence of marketing has two elements – developing products that customers want and persuading people to buy them. The 'people' here come in layers, for there is usually a merchant of some kind between the manufacturer and the ultimate customer and often a 'supply chain' of successive buyers: international buyers who buy tea, say, from the growers, shippers who arrange transport to wholesale markets around the world, wholesale buyers who blend and package it for sale to shops, which sell it to us. (As a result, the tea grower gets only a small fraction of the ultimate price; those who tend and pick the tea much less.)

At each stage there is a market operating in which the sellers at that stage compete with each other on quality and price. Persuasion backed by data is at the heart of this: convincing the buyer that 'our product' will make you more money than 'theirs'. Manufacturers spend money on gathering data to support their case – though not nearly as much as on advertising it! These principles apply to products of all kinds, from cabbages to prescription drugs.

Price setting

Marketers decide on a selling price in the context of 'the 4 Ps' – product, price, place, and promotion – alongside an understanding of the market and the 'positioning' of the brand within it. For basic products, like those sold in supermarkets, 'market and brand analysis' takes into account many variables:

- The current portfolio of our products and prices.
- Where this product might best fit in the market, where we price it, and why.
- Comparing it to key competitors' similar products.
- Looking for a unique benefit or added value to warrant a premium.
- Consumer research – what consumers are willing to pay, any 'price ceiling' or not.
- Where it will be sold, and if there might be a different pricing approach by channel – for supermarkets, convenience stores, or wholesale channels, for example.

The actual costs in producing, delivering, and advertising the goods are, of course, a significant factor – as is the company's expectation of profit margin. But the impact of a product's contribution to the total brand profitability is a key consideration. Even if the product does not make money, a company may launch it as playing an essential role for the brand – perhaps to drive volume and operational efficiencies, to satisfy a loyal consumer base, or to protect market share. Some products may be sold below cost (or at least at a reduced profit margin) as 'loss leaders' to attract customers to the brand. Others, especially 'extras' sold after the customer has chosen the main product, may be sold at a huge markup to compensate for a low initial cost. In summary, marketers take a broad view based on the total brand profit and loss. Marketing clearly involves a diverse range of modelling competencies that characterise mathematical literacy in linking human understanding with analytical and mathematical skills.

For premium and luxury products, consumers buying decisions are not primarily based on price. The marketing approach has the same elements but brand strength, its role and positioning in the market are key factors. If you think of, say, Apple, consumers want to own these products because they buy into the 'Apple vision' and design ethos, and/or have prior experience. Price is probably low down in the 'consumer purchase decision tree'. Apple can use this in their pricing models through mark ups on their products – how big we can only speculate but the profitability of the company gives a clue.

Similar principles apply to services, though for these the supply chain is shorter, often only one link – as with your hairdresser or lawyer, for example. But medical services, for example, can have quite complex chains – from primary care physician (or ambulance) to a hospital with its diverse range of expertise deployed in variously structured ways.

A consumer decision

The following activities reflect consumer decisions that will be familiar to students. They will also serve to stimulate further reflection after the later sections of this chapter.

Choosing a mobile phone is something of a cliché here, but smartphone ownership is ubiquitous around the world, and making a good choice is complicated by the way phones are sold, often bundled with finance plans and service subscriptions. When you have chosen the actual phone and level of service you want, even a single vendor will typically offer multiple plans for the same phone. Then you may have a choice of vendors, or even the option of buying the hardware and service from separate suppliers. Each service provider usually offers a range of "tariffs" offering different permutations of call options, data limits, and contract lengths. The concepts of making quantitative estimates of your actual needs, interpreting the marketing, working out the total cost of ownership, choosing valid measures for making comparisons, and interpreting the results make this a powerful example of informed decision-making. In some countries, where you live might force you to choose a particular network, restricting the choice somewhat, but in other areas (such as the UK and EU) there is a fairly competitive mobile phone market that gives most people a choice of several networks and service providers offering options at very different prices.

For these activities we'll assume that the hardware marketing folk have done their job well and everybody wants the amazing nuPhone 42b more than their next hot meal (actually choosing between phones is a whole other adventure, and any discussion of that will be out of date before this book is printed). So, the question is, how do you pay for it and what service subscription do you want?

ACTIVITY

Students discuss and research what level of phone service they need.

How many minutes of phone calls do you make in a month? Are any of them abroad? How can you estimate this? Is making traditional phone calls and sending texts even important to you – if not, how many gigabytes of data do you need to download in a month? If you already have a phone, then you probably have information on past usage.

Discuss the following example. A phone plan offers 125 GB of data per month at speeds of up to 100 Mbps. (Remember to think carefully about units!)

- How many hours per month could you spend downloading at top speed? Is this realistic?
- How many hours of music could you download per month?
- How many hours of video could you download per month?

Use the internet to look at some phone service plans (SIM-only contracts) and see what choices there are.

It is probably simplest to concentrate on 12–24-month service-only contracts at this stage to ensure like-for-like comparisons. If you choose "pay as you go" schemes that typically have a time limit, often 30 days, you could consider including monthly top-ups to have the equivalent of a contract. The issue of contracts versus pay-as-you-go and the implications of making a contractual commitment can be reintroduced after the second activity, which introduces the total cost of ownership.

ACTIVITY

Students research the cost of a phone over several years.

Now you have an idea what sort of phone plan you need, how much will it cost, in the long term, to pay for the phone and service? Collect data from the internet and build a spreadsheet to record it in an organised way. Questions you might want to ask are as follows:

- How much do you have to pay up front, and how much per month?
- How long are you locked-in to the contract? What is the total you will pay in this time? Are you likely to keep the phone for longer?
- Can you save any money by paying more up front? How does this depend on how long you keep the phone?
- Can you buy the phone more cheaply elsewhere and just get the service from the phone company?

Deal	Seller	Cost Up front	Cost Monthly	Months (contract)	Total cost after N months 12	24	36	48
Phone & service on 1 year contract	PhonesCo	25	69	12	853	1681	2509	3337
Phone & service on 2 year contract	PhonesCo	25	47	24	589	1153	1717	2281
Phone & service on 2 year contract	UPhoneMe	10	44	24	538	1066	1594	2122
Phone & service on 3 year contract	MegaFone	30	47	36	594	1158	1722	2286
Phone & service on 3 year contract	PhonesCo	25	41	36	517	1009	1501	1993
Buy phone up front from PhonesCo	PhonesCo	500	26	36	812	1124	1436	1748
Phone from BuyMart, 24 month service	UPhoneMe	550	28	24	886	1222	1558	1894
Phone from BuyMart, 24 month service	PhonesCo	550	24	24	838	1126	1414	1702
Phone from BuyMart, Per-month service	PhonesCo	550	30	1	910	1270	1630	1990
New plan for existing paid-off phone	FoneNet	0	26	36	312	624	936	1248
New phone on your existing plan	BuyMore	550	18	12	766	982	1198	1414

Means contract is over and you can shop around for better service

FIGURE 7.1.2 Spreadsheet to compare total cost of phone ownership.

Source: Spreadsheet available online [7D].

There is unlikely to be a single clear "right" answer – you will need to think about how long you want to commit yourself for, how much you can afford to pay up front. The important thing is to be aware of the 'total cost of ownership' (how much you will spend on the phone over its lifetime) in various scenarios. Figure 7.1.2 shows an example spreadsheet, *but* it is important to look up current data and possibly adapt the format of the spreadsheet, as the way phones are marketed varies with time and between countries. In recent years, some phone companies have made it clearer how much you are paying for the phone and how much for the service – bundling them together (as shown in the example) was common in the past, when commonly the phone company continued to keep charging the same monthly rate long after the contract period had ended and the cost of the phone has been paid off. Sometimes it is more expensive to get the phone from an electronics store than a phone company – who make their money back on the service charges – but that could still work out cheaper if you keep the phone for longer and can shop around for a cheaper service contract.

7.2 Advertising

Many Western countries have laws that limit the exaggeration that is permitted in advertising. In the UK, the Advertising Standards Agency has codes for various areas of advertising (print and broadcasting in particular). At the top level, it states:

> The central principle for all marketing communications is that they should be legal, decent, honest and truthful. All marketing communications should be prepared with a sense of responsibility to consumers and society and should reflect the spirit, not merely the letter, of the Code.

There is a lot of detailed guidance that marketers have to follow, more or less, as to what this means in different areas of advertising. Not surprisingly, many advertisers test the boundaries of the regulations – for which some are fined or simply forced to withdraw the ad.

The internet still retains something of the Wild West on advertising, as for so much else. The central principle set out in the UK ASA Code and its international equivalents remains the goal, but the challenge of turning it into workable regulations remains elusive – and, to some, undesirable. Most of the focus of regulation was initially on avoiding harm, particularly to children, and on protecting an individual's data from commercial exploitation without their consent.

In the United States, the guarantee of freedom of speech – a precious freedom that is specifically enshrined in the Bill of Rights as the First Amendment to the Constitution – is sometimes used as a defence against accusations of misleading advertising, which makes the regulation of advertising more difficult.

Within whatever constraints may apply, advertisers have devised many modes of persuasion, often using numbers to reinforce their message. At the naive end, we have claims in the style of "8 out of 10 cats who expressed a preference said Wondercreme increased the length of their whiskers". The manufacturer will almost certainly be able to produce the results of a survey that *appears* to show this – whether it would pass any test of peer review or reproducibility is another matter. This kind of advertisement shows how you can make a perfectly truthful statement about a 'survey' while falsely implying that something is scientifically proven. "Probably the best lager in the world" is in this class, without even bothering with numbers.

The adverts in Figure 7.2.1 are imaginary, but they parody widely seen advertising tricks. Simply quoting an irrelevant fact can make your bottled

FIGURE 7.2.1 Parodies of advertisement tropes.

water sound healthier than the equally fat-free stuff that comes out of the tap. Few 'miracle' diet or exercise plans – even ones with some validity – will work without additional major lifestyle changes ('as part of your calory controlled diet' is the usual disclaimer). The third example, which arguably crosses the line from 'clever advertising' to 'outright scam' is based on a story from long ago (that might even be true) about a paid advert in a newspaper – so the scammer took the risk that not enough people would bite to recoup the costs. Today, mass distribution via social media or junk email costs virtually nothing, so even a tiny 'success' rate can be profitable.

ACTIVITY

Questionable ads.

Ask students to bring in examples of advertisements they think are misleading or might be misleading to some people, particularly if they rely on numbers. Small groups then discuss and classify the methods used to mislead.

Once a marketer has attracted a customer's interest with a well-tailored advertisement – they're in the shop (or on the website) – there are further tactics to seal the deal, stave off competition, and maximise profit.

The price on the sticker is the key number the customer sees, and their first impression of how expensive the product is. Too high and it will scare off customers. Too low and the seller is sacrificing profit. Techniques for minimising the advertised price and obscuring the total cost of ownership include the following:

- 'Subscription services' paying monthly versus lump sum one off purchases have long been common for expensive things like cars ("Only $600 a month for 3 years"). This approach is now common for phone apps too ("Only $4.99" – 'per month' is implied). It is spreading to physical goods – for heated seats in your new car, for example – as a way to lower the purchase price, while also providing a steady income stream for the manufacturer.
- 'The King Camp Gillette' business model: give away the razors and then charge a fortune for the blades. More modern examples include cheap printers with expensive ink, or cheap coffee makers that use proprietary coffee pods.
- 'Optional' extras: the advertised price gets you the bare-bones model, but maybe you'll regret not getting the more comfortable saddle for your bike,

the bigger battery for your electric car, more memory, and storage for your phone/computer. The trick for the seller is to sell you these *after* you've decided to buy the product at the sticker price.

In this kind of situation, taking time out to do simple calculations will usually let you compare the up front, annual, and total lifetime costs, leaving you to consider which is more important to you at the time. If, for example, you think that $200 of extra storage will make your new computer last 4 years rather than 3 years, how will that change the annual cost?

Another way of making money beyond the listed price is to sell the customer an *extended warranty* to insure against needing expensive repairs in the future. As we asked in the Introduction, is it likely to save money? Here it is important to consider when any breakdowns are most likely to occur. Usually, this will follow the pattern of the 'bathtub curve' (see [7E] for background) shown in Figure 7.2.2.

The initial period when manufacturing faults show up is covered by the standard guarantee. Wear and tear faults from repeated use increase after many years, normally beyond the period of the extended warranty where the likelihood of breakdown is at its minimum. The generosity of the standard/statutory warranty varies between countries, and some extended warranties will include cover for accidental damage that wouldn't be covered otherwise but, in general, it is more expensive to insure against risks that you can comfortably afford to cover than to pay when problems arise – insurance companies make profits. This is another area in which bit of personal accounting and modelling can help: how much do you typically spend, over a few years, on repairing/replacing broken devices? How much would it cost to pay for extended warranties on everything you buy?

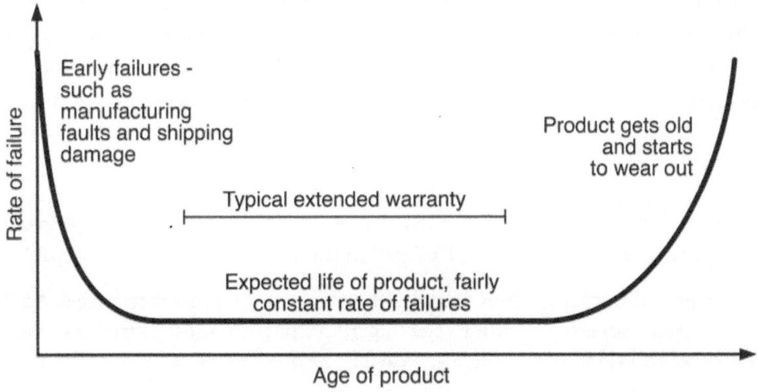

FIGURE 7.2.2 Likelihood of breakdown of a typical consumer product.

7.3 Selling in politics

While most countries have laws against false or misleading advertising of products and services, these rarely extend to claims and promises made by politicians, who are free to mislead in whatever way they think they can get away with. It is against the law for a manufacturer to falsely claim their car will do 100 km on 1 litre of petrol, but not for a politician to say that Britain's economic performance over the last 10 years has been the best among the OECD countries.

This situation is primarily justified as freedom of speech. It runs the same risk of refutation by trusted fact-checkers with the associated loss of credibility and trust.

Claims for the future are also excused, sometimes even excusable, because of the general unreliability of economic forecasting – a field described by one distinguished economist as "designed to give astrology a good name". Nonetheless, evaluating political statements presents an ongoing challenge for the mathematically literate in a very important domain.

In the UK, the major broadcasters (BBC, ITV, and Channel 4) have a remit to present news and current affairs in an unbiased way that includes opposing views – one which they proudly claim to uphold. Print media, including newspapers, have no such constraints. Nonetheless, the 'leading papers' work hard to achieve a reputation for presenting objective truth by separating news reporting from opinion, following the (Manchester) *Guardian*'s long-time editor, C.P. Scott (1921):

> Comment is free but facts are sacred.

News media – originally newspapers but increasingly moving online – in the United States that embrace this principle include *The New York Times* and *The Washington Post*, *The Age*, and the *Sydney Morning Herald* in Australia. We shall look at three UK publications. While the political viewpoint of each is widely recognised, and clear in their comment sections, these organisations claim to present their news reporting as objective and unbiased. While broadly true, there is some 'leakage' of opinion into news. Figure 7.3.1 shows

The Guardian 15 April 2023	*The Times* 15 April 2023	*The Telegraph* 15 April 2023
Further England nurses' strikes present 'severe challenge' to NHS	**Nurses call 48-hour strike after rejecting RCN pay deal**	**Nurses to target A&E in most extreme strike yet**

FIGURE 7.3.1 UK news headlines – same news, different slant.

corresponding headlines from the three sources (one regarded as 'progressive', two as 'conservative') in a dispute reflecting offers of pay increases to nurses that are well below price inflation.

Note that *The Guardian* presents the strikes as a challenge for the healthcare system for which the (Conservative) government is responsible; *The Telegraph* calls the strike 'extreme' and 'targeted' on emergency care, while *The Times* suggests the offer comes from the trade union – which did indeed recommend its members to accept the government's offer as the best that could be achieved. Other news sources make no attempt to separate news from opinion.

This 'leakage' of opinion into news reporting, often far more blatant than Figure 7.3.1, has led some to claim that there is no such thing as objective truth. The historical record, with its inevitable selection of facts, offers evidence to support this view – the inclusion, or not, in history curricula of slavery and its role in enabling Western economic success being a prime example. Nonetheless, for literacy, mathematical, and, in general, to survive and thrive, a search for objective truth through analysis of data and critical thinking is central.

Ways of 'improving' an argument

While there is no reliable way of predicting the future in politics or economics, the rhetoric used by political parties for selling policies often relies on the following:

- An account of past performance, which is verifiable though inevitably complex.
- Comparisons between periods and/or between countries.

What are the essential components of these arguments and what ways of misleading the listener or reader are common?

The choice of variables is, as always, crucial. For example, the favourite indicator of the overall performance of an economy is GDP, the 'gross domestic product' or, closely similar, the national income. However, since this says nothing about where this income goes, it is a poor indicator from a personal point of view. For most people the *real median household income*, adjusted for inflation, is a more significant indicator of economic progress. (Avoid the *mean* household income, which goes up even if all the benefit of an increasing GDP goes to the very rich.)

Details in the selection of data can crucially strengthen or weaken a case – for example, by choosing different start and end dates for the data used to calculate summary statistics. This is illustrated by Figure 7.3.2, which shows UK GDP from 1992 to 2021.

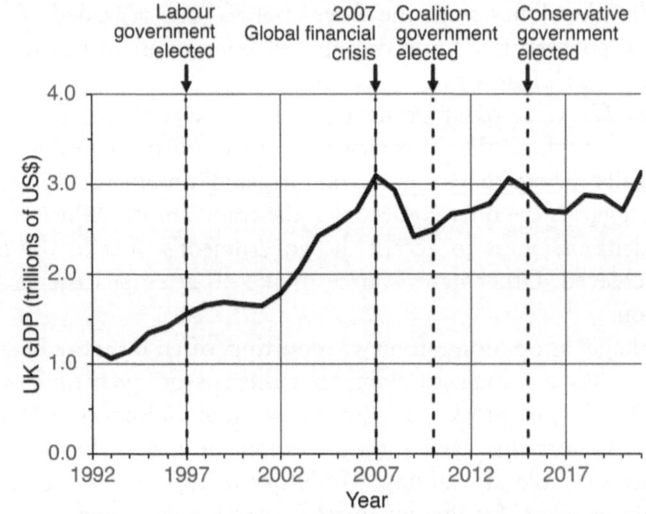

FIGURE 7.3.2 UK GDP 1992–2021 showing key events.

Source: Using open data from The World Bank: World Development Indicators 1/3/2023. [7F]. Often used for international comparisons.

> **ACTIVITY**
>
> GDP growth
>
> Calculate the annual GDP growth rate that, if compounded over 10 years, would produce GDP growth by a factor of about 2. Remember that growth is measured in multiplicative terms (rates, ratios, percents), rather like interest (see Chapter 9). This can be done using logarithms.
>
> Alternatively, a spreadsheet can be used to quickly converge on the answer using a guess-check–improve approach.

The appropriate formula for annual percentage rate R is as follows:

$$R = 100\left(e^{\left(\frac{\log_e G}{P}\right)} - 1\right), \text{ where } G = \frac{\text{end GDP}}{\text{start GDP}} \text{ and } P = \text{period in years}$$

For the growth factor of 2 over 10 years, this gives an annual compound growth rate of 7.2% (to check calculate 1.072^{10}). These figures match growth in UK GDP from 1997, when a Labour government took over until the global financial crisis of 2008. Those of a more conservative view would correctly point out that by 2010, when the government changed, the growth since 1997 averaged only 3.7% (a factor of 1.6 over 13 years). In rebuttal, it could be pointed out that for 2010–2021, the growth rate was only 2.1% (1.25 over 11 years) with no growth at all since 2015. These are very different economic and political stories. Note that the graph provides a far more informative and detailed picture of the situation than the summary statistics we just noted.

Playing with units and rates is another common device for giving a more favourable impression – announcing the amount of new funding "over the next three years" or "over the lifetime of this administration" (4–5 years, depending on country) rather than on a standard annual basis makes the government look more generous, until you think about it.

Reissuing announcements of funding that had already been promised previously is another common device. Sometimes this involves redirecting it to another area. 'Robbing Peter to pay Paul' in this way is common.

Antidotes to distortions

The first line of defence against misleading political arguments is the media – traditionally, the press and TV broadcasters. This has been disrupted by the rise of the internet, where the distinction between 'news' and 'social media' commentary becomes blurred. Countries which genuinely aspire to having a 'free press' generally have a range of news providers with political analysts who dissect the statements of government and opposition politicians. As we have noted, most of them have a well-recognised political viewpoint, more or less extreme, that needs to be taken into account. There is an even wider range of choices on the internet, but only a minority aim for unbiased analysis of political statements. The rest amounts to a flood of opinion, often very strongly held and backed up by dis-information. So, again, in order to get a reasonably objective overview, it is important to read the analysis in a range of sources, including some that you regard as biased.

Some broadcasting organisations (including the BBC, ITV, and Channel 4 in the United Kingdom and NPR in the United States) have a public service remit that requires them to hold politicians of all persuasions to account, asking the awkward questions. The answers – or frequently, the evasions to avoid answering directly – give the listener some idea of where the weaknesses lie. (That all political parties criticise the BBC as biased against them is some indication that it handles the public service remit pretty well.) This can, however, lead to 'bias by balance' wherein a well-intentioned remit to include a range of voices can result in minority views appearing more widely held than they really are. For a long time 'climate change deniers' succeeded in this way – for example, the BBC would give equal time to a leading climate scientist and a retired politician who claimed the overwhelming evidence was inconclusive.

In the end a rigorous analysis of a political argument means going back to trusted primary sources. Most countries have national statistical agencies that not only produce reliable data on important areas but are also responsible for vetting government data to ensure that, while it may be selective, it is not actively misleading.

While the majority of 'think tanks' have a political agenda – well recognised if not admitted – there are some that have built a reputation for objectivity without bias. The UK Institute for Fiscal Studies is an example – an enviable reputation to achieve.

Confirmation bias and common misconceptions

It is important to remember that we all are, at least partially prone to 'confirmation bias', the 'echo chamber' effect – we seek sources that we are likely to agree with, which provide comforting confirmation, particularly in areas where we hold strong opinions. (The choice of examples in this chapter doubtless reflects this.) To minimise this effect, it is important for informative discussions to have people with opposing views. Personal vigilance is the only antidote.

In many cases, people's strongly held misconceptions have been influenced by popular culture – non-factual 'memes' that permeate news and entertainment media, politics, and 'what the bloke down the pub said'. Exploring and analysing real-world data is one way of challenging this. The *Gapminder* website [7G] offers several tools to support this. The headline feature is the *Worldview upgrader* – a quiz with questions such as the one shown in Figure 7.3.3 that play on widely held misconceptions. It won't take long for a smart person to spot the pattern of 'common knowledge' being wrong, but this is justified by the survey data presented after each answer showing

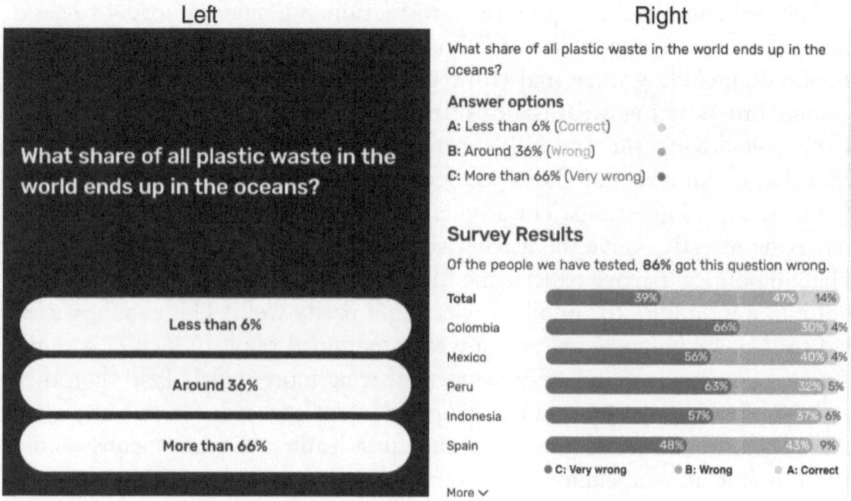

FIGURE 7.3.3 Example question from *Worldview upgrader* and follow-up.

Source: Screenshot from Gapminder [7G] – modified to add key for black and white version.

how widely held the counterfactual views are. If using the quiz, it is important to pay attention to this follow-up information, especially the part that explains why the misconceptions can cause problems, even when the truth seems more optimistic. *Gapminder* also offers a range of animated, interactive graphs of demographic data, and supporting videos, which can be used to explore issues raised by the quiz – these are highly interactive and colourful, and best experienced by visiting the website. Sources of online data and visualisation tools are discussed further in Chapter 10.

Some social media platforms exploit confirmation bias by targeting users with material based on their previous browsing habits. In many ways, social media have democratised the media, allowing people to share their views without 'the establishment' – politicians, newspapers, and broadcasters – acting as gatekeepers, curators, and censors. However, the resulting flood of information and misinformation is too much for most people to sift through, so the 'new gatekeepers' are the algorithms which recommend new material to users based on their past browsing habits and other harvested personal information. These algorithms are a crucial part of the big social media platforms and – at best, assuming no more sinister motive – are designed with the sole intention of keeping users on that platform by feeding them more of what they like – there is no incentive to risk alienating someone by challenging their views. In any case, these algorithms work automatically without thought or judgement. While this is also true of old-fashioned printed newspapers, at least their biases are well known and their sometimes-conflicting headlines visible to anybody browsing the newsagent's shelves – they have at least some incentive to be seen as honest, reputable journalists. The social media giants are disinclined to take decisive action against misinformation not because they have some evil plan, but because their entire business model relies on automatically processing vast quantities of material contributed by customers, and any meaningful level of human curation would be prohibitively expensive. They say they are not pretending to be journalists – they are simply connecting you with your newly expanded circle of 'friends'.

ACTIVITY

Discuss how social media have affected the informative discussion of issues.

On bullshit

We noted earlier that there is an approach to persuasion that doesn't worry about the consequences of telling lies. The method is to bury the listener or reader in bullshit – a string of assertions, sounding more or less plausible, that support whatever argument the speaker wants to make. Truth is irrelevant. In Frankfurt (2005), the Princeton philosopher gives a readable account of bullshit and how to spot it. Long familiar in social groups with drinks in hand, in recent years some charismatic politicians have found that it works to advance their careers, at least for a while. So recognising the difference between lies and bullshit has become an essential skill for assessing 'information', in our politics as elsewhere. Being mathematically literate can help.

Acknowledgements

Figure 7.3.2 was created using open data from The World Bank: World Development Indicators 1/3/2023. [7F].

Figure 7.3.3 is a screenshot from the *World View Updater* website – free material from www.gapminder.org [7G].

References

Frankfurt, H. G. (2005). *On bullshit.* Princeton University Press.

Links to useful material

To visit any of these links, scan this QR code or visit ltml.mathlit.org – append the link code to go directly to the entry – for example, ltml.mathlit.org/7A

The original source links are given below for attribution purposes:

[7A] *The Secret World of Fake Online Reviews* – The Guardian (22 April 2023)
https://www.theguardian.com/money/2023/apr/22/it-can-be-incredibly-profitable-the-secret-world-of-fake-online-reviews

[7B] *Best Washing Machines to Buy in Australia* – Choice
https://www.choice.com.au/home-and-living/laundry-and-cleaning/washing-machines/review-and-compare/washing-machines

[7C] *Decathlon: The Art of Scoring Points* – NRICH
https://nrich.maths.org/8346

[7D] *Spreadsheet to Compare Total Cost of Phone Ownership* – mathlit.org
https://ltml.mathlit.org/7D

[7E] *Bathtub Curve* – Wikipedia
https://en.wikipedia.org/wiki/Bathtub_curve

[7F] *World Development Indicators* – The World Bank
https://databank.worldbank.org/source/world-development-indicators

[7G] *Worldview Upgrader* – Gapminder
https://www.gapminder.org/

8

EQUALITY AND INEQUALITY

Inequality of wealth and income is a topic that excites the interest of people, particularly young people whose relative position has declined in recent decades. As the word implies, it is about comparisons between different groups at different times, offering an opportunity for discussions of important social and political issues where numbers are central. But, as with any complex field, there are many issues to consider. When, for example, should we include wealth, where the disparities are even larger, along with income? Finding appropriate data and choosing it to tell a story is at the heart of mathematical literacy. Here more than in most areas, the choice a person makes in presenting a case will depend on the message they want to get across – which is interesting in itself. So this is very much a 'know about' chapter, where critical thinking is central.

For example, a popular measure of inequality is the 'wage ratio' between the chief executives of large companies and the median wage of their workers. In the United States, this rose from about 20 in 1950 through 120 in 2000 to 650 in 2022. An analysis noted that "on the first workday of 2023, CEOs will make more than the average annual pay for all US workers". (*Quick check:* If the average working year is 260 days per year, the wage ratio needs to be 260 for the first day pay to equal average annual pay.) The ratio has increased over the last 50 years in other countries, too, from single digits to around 100 in the UK, in Sweden to 50. There is no evidence that this reflects improved company performance.

DOI: 10.4324/9781003303503-9

ACTIVITY

Is inequality 'a bad thing'? If so, in what ways. Small groups discuss the advantages and disadvantage of inequality of income, as represented by the ratio of CEO to median worker income. What ratio seems appropriate?

This data represents a surprisingly extreme contrast in incomes, yet comparisons of wealth yield larger ratios. Is there a way to get a picture that is both broader and more balanced so as to inform discussions – in the classroom or elsewhere?

8.1 A historical analysis of wealth and income

The research on these issues by the distinguished economic historian Thomas Piketty is summarised in his book *A Brief History of Equality* (Piketty, 2022) – a very readable account for the non-expert mathematically literate. Here we outline some of the issues by presenting just a few of the comparative graphs from the book – all are available from Thomas Piketty's website [8A]. This website offers a rich source of information for small group and class discussions of the many important issues involved.

Figure 8.1.1 shows how both the total wealth of the world and the population have grown over the last 300 years. (They were relatively stagnant through the previous centuries.)

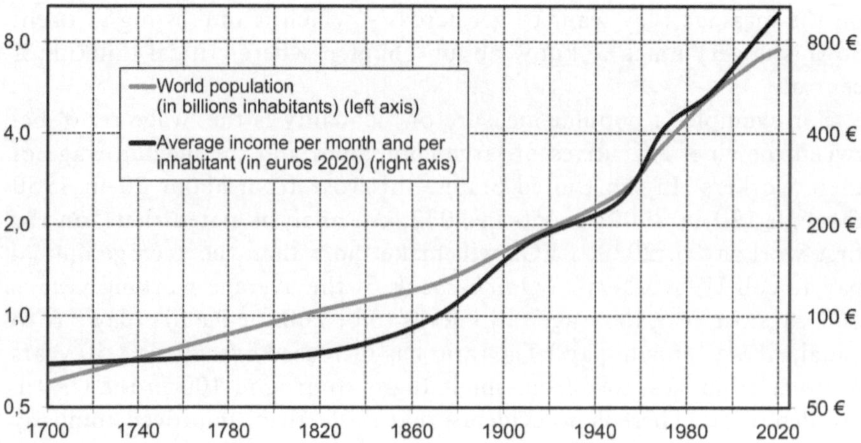

FIGURE 8.1.1 World population and average income 1700–2020.

Source: Adapted from Piketty (2022) [8A].

Briefly, world population and average national income per person both increased more than tenfold between 1700 and 2020 – from about 600 million inhabitants in 1700 to over 7 billion in 2020, while income per person (expressed in 2020 euros with 'purchasing power parity') increased from barely €80 per month per person in 1700 to €1000 per month per person in 2020. This corresponds in both cases to an average annual growth rate of about 0.8%, compounded over 320 years. Note this means that total world income (GDP) increased at an average rate of about 1.6%. (*Mathematical literacy question:* 0.8% more people each year each with 0.8% more income – why is adding the rates reasonable here?)

So we have got richer. But is it really that simple? Does '2020 purchasing power parity' mean anything before the industrial revolution, or at a time when you could go to prison for being in debt. Piketty (2022) outlines how those issues have been investigated in a comprehensive programme of historical analysis of prices, wages, wealth, and other variables. Equally, he recognises that in comparing such different 'worlds', the results are only qualitatively meaningful. However, his conclusion that the average person is economically much better off than 300 years ago is surely plausible.

Average income can be a useful measure for comparisons between countries and different periods, but it obscures inequality between people at different levels of income *within* each country. Piketty often tracks the income and the wealth of three groups: the top 10%, the middle 40%, and the bottom 50%, comparing them over time and relating them to political and economic changes in the periods concerned. He observes that, while all groups have become more prosperous, the general trend is for increases in prosperity to go mainly to the already wealthy – a trend that has only been contained by active government intervention, mainly through redistributive tax systems.

This issue is vividly illustrated by comparing the post–World War II 'egalitarian' period 1945–1980 with the more individualistic 'neoliberal' economics that was introduced in the 1980s by Fraser (Australia), Thatcher (UK), and Reagan (USA). This approach is built on the economic ideas of Hayek, Friedman, and others in the 'University of Chicago School'. They argued that individual entrepreneurship aided by low tax rates would produce faster growth (GDP) whose benefits would "trickle down" to raise the incomes of everybody. "All boats rise together with the tide of increasing prosperity". It has not turned out that way.

Figures 8.1.2 and 8.1.3 show how the sharing of wealth and income between the classes changed between 1900 and 2020. In Europe as in the United States, we see that between 1914 and 1980 a steep decline in the share of the richest 10% in total private property (real estate, business and financial assets, net of debt) to the benefit principally of the middle 40%. This movement is then partially reversed between 1980 and 2020, notably

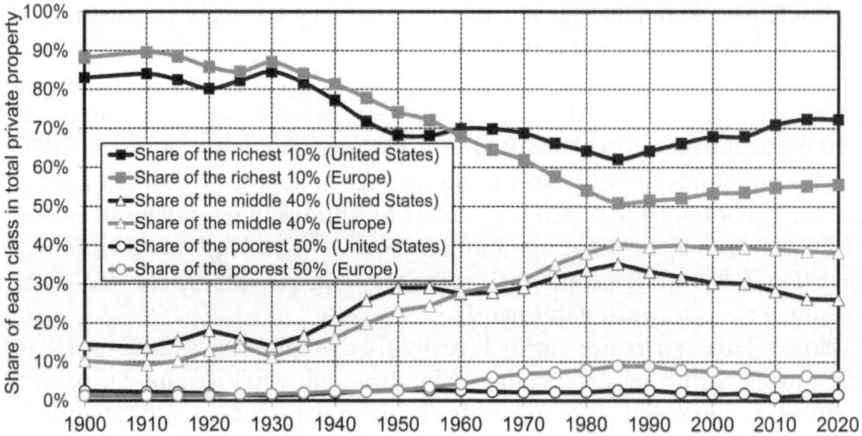

FIGURE 8.1.2 Wealth in Europe and the United States, 1900–2020: The birth and fragility of a patrimonial middle class.

Source: Adapted from Piketty (2022) [8A].

in the United States. ("Europe" here is an average of the data from France, Germany, Sweden, and Britain.) Note that these percent shares are not per person. In Europe in 2020, for example, the top 10% together had well more than 8 times the wealth of all the people in the bottom 50% together, so individuals at the top had more than 40 times the wealth of individuals at the bottom. In the United States, the situation, as the graph shows, the ratio is more extreme.

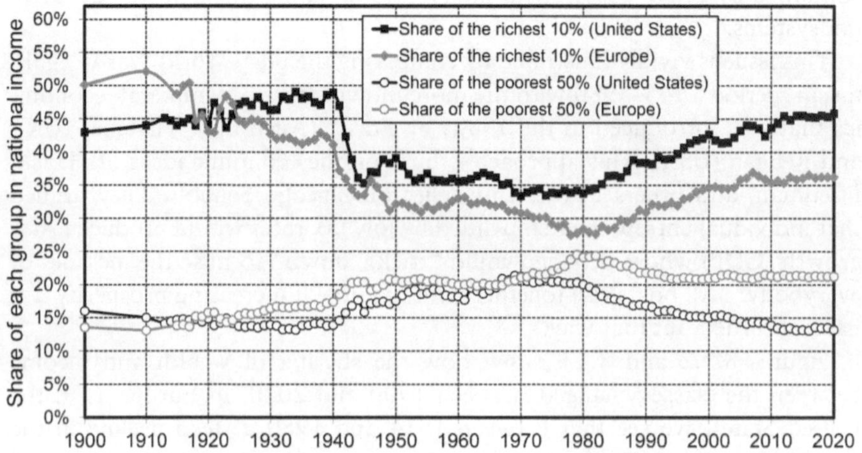

FIGURE 8.1.3 Income inequality: Europe and the United States, 1900–2020.

Source: Adapted from Piketty (2022) [8A].

Figure 8.1.3 shows the share in national income of the richest 10% versus the poorest 50% and how the pattern of income changed over the same period. In Europe, income inequality has started to rise again since 1980, although remaining at levels clearly lower than those of 1900–1910. The increase in inequality has been much greater in the United States. In both cases, inequality has remained high: the richest 10%, though five times fewer, still receive a share of total income much larger than the poorest 50% receive – on average about 15 times greater per person in the United States, 8 times in Europe, with huge disparities between individuals in each case, particularly within the top 10%.

How far does this reflect the pattern of taxation?

Here we focus on the United States where, from 1915 to 1980, the tax system was highly progressive, in the sense that the effective tax rate paid by the highest income groups (taxes of every type included, then expressed as a percentage of pre-tax income) was substantially larger than the average effective tax rate paid by the total population (and particularly by the bottom 50% incomes). Since 1980, the tax system has been much less progressive, with little differences in effective tax rates across groups, as shown in Figure 8.1.4. In 2020, people in all groups paid about 30% of their income in taxes.

However, the promised resurgence of growth following the cut in top tax rates, predicted by the neoliberal economists and their often-wealthy supporters, did not occur. In the United States, the top marginal tax rate applied to the highest incomes dropped from 72% to 35% and the growth rate of per capita national income *dropped* from 2.2% to 1.1% per year over the

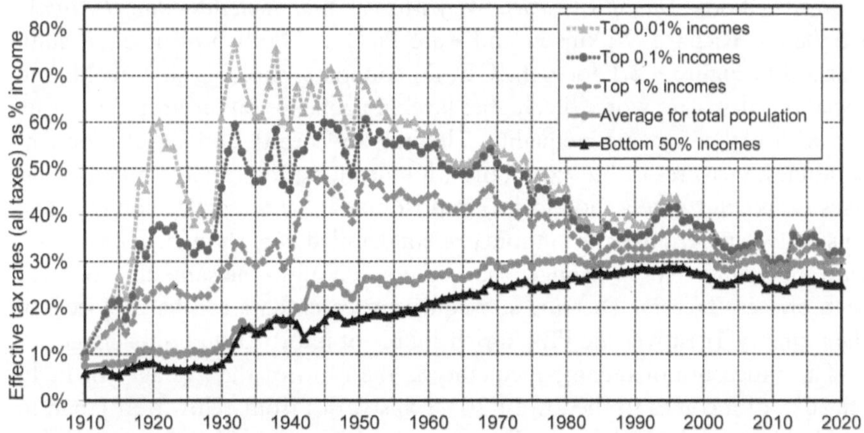

FIGURE 8.1.4 Effective tax rates and progressivity in the United States 1910–2020.
Source: Adapted from Piketty (2022) [8A].

TABLE 8.1.5 Growth and progressive taxation in the United States, 1870–2020

Period	*1870–1910*	*1910–1950*	*1950–1990*	*1990–2020*
Top tax rate	0.8%	55%	72%	35%
Growth rate	1.8%	2.1%	2.2%	1.1%

Source: Data from Piketty (2022) [8A].

same period – as shown in Table 8.1.5. And, as we saw above, far from the predicted trickling down of riches so that "all boats rise together with the tide", the inequality of distribution of wealth and income surged upwards. There were similar but less extreme effects in Europe, where the diverse tax systems make comparisons more complicated.

ACTIVITY

A critical discussion of the pattern of wealth and income data from this section. Students can be encouraged to look for other data, from piketty.pse.ens.fr/equality and other sources, to be used in their preparation.

8.2 Social effects of inequality

What are the effects of inequality? Is inequality 'a bad thing'? After all it is broadly favoured by the wealthier, who don't like to pay a lot of tax (who does?), and generally accepted as a fact of life by those who have little but 'somehow get by'.

In their book *The Spirit Level: Why More Equal Societies Almost Always Do Better,* Richard Wilkinson and Kate Pickett (2009) present data, summarised in Figure 8.2.1 for rich (OECD) countries, showing that health and social problems are worse (i.e. higher level of problems on the graph) in societies with higher income inequality. The index of health and social problems combines measures of life expectancy, infant mortality, teenage pregnancies, obesity, mental illness, homicides, imprisonment rates, social mobility, and trust between citizens. The authors present detailed data showing correlations with inequality – some stronger than others, but all in the same direction for each of these factors. This and subsequent research data can be explored on the Equality Trust website [8B], a rich source of information on this issue.

Their measure of income inequality is the ratio of the averages of highest 20% of incomes to the lowest 20%; this varies from below 4 in Japan to nearly 10 in the United States. They also show that the index does *not* correlate with the *average income* of the country – Greece and Ireland, for example, are close in income inequality despite a 50% difference in average income.

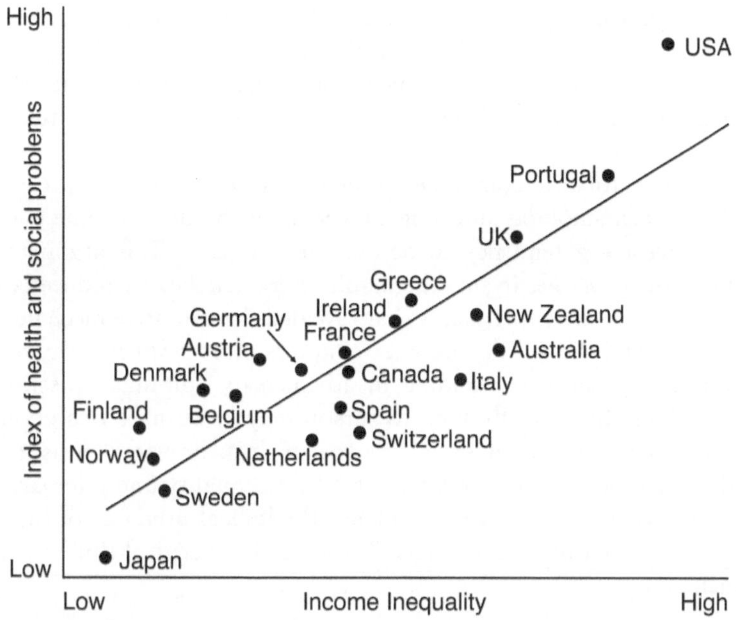

FIGURE 8.2.1 Index of health and social problems versus income inequality.

Source: Adapted from Wilkinson and Pickett (2009) [8B].

> **ACTIVITY**
>
> Use the data from *The Spirit Level* [8B] to inform discussions of the social effects of income inequality on different income groups. Ask each student to choose one graph and explain its implications for discussion in the group.

> **ACTIVITY**
>
> Examine the differences between the measures of income inequality that are used by Wilkinson and Pickett (e.g. Figure 8.2.1) and Piketty (Figure 8.1.3) and discuss strengths and weaknesses.

Methodological issues

In considering this data and the dramatic inferences that it suggests, there are some important issues of principle that a mathematically literate person may well be considering. The data shows correlations between measures of inequality and social outcomes – but correlation does not necessarily imply causation, nor say in which direction that might flow. Since it is normally impossible to do controlled experiments on societal variables, one relies on observing the outcomes of *changes* that arise naturally or through the effects of policy decisions.

For example, tobacco companies argued for a long time that the clear correlation between smoking and lung cancer might be due to some common causative factor – a tendency to be anxious, perhaps. This argument lost credibility when *changes* in smoking habits were matched by reduced cancer deaths – as well as by experiments that revealed the causative mechanisms.

Wilkinson and Pickett devote later chapters of *The Spirit Level* to such methodological issues, responding in advance to legitimate questions of this kind. A mathematically literate person may have questions about the mathematical modelling of social constructs (e.g. how the measures are created – for more on this see Chapter 12), the quality and comparability of data over time and between countries, the logical premises of the arguments (e.g. correlation/causation as above), and the critical thinking about the conclusions.

8.3 Approaches to reducing inequality

The Spirit Level also looks at the springs of inequality in different societies and the kinds of change in policy and practice that might reduce it. They point, for example, to the potential of 'pre-distribution' through reducing inequality of income (as in Japan) rather than relying on 'redistribution' through taxation. They advocate a range of 'levers' to reduce inequality, including more progressive taxation as in the past, a change in corporate law from its focus on shareholder value to include other stakeholders (e.g. leading to better paid jobs for employees), and a social 'safety net' that supports a decent standard of living.

Complementing this broad look at various mechanisms, Piketty describes a specific approach. While his historical analysis has been widely accepted, his proposed way forward is much more controversial. Nevertheless, it is a useful starting point for discussions of the important issues raised.

Piketty's model for *participatory socialism* treats wealth and income as comparably important. It is based on redistribution through two channels: the circulation of property between generations – a *minimum inheritance* to be allocated to each young adult at 25 years of age – and a *basic income* for all. The effect of the former is illustrated in Figure 8.3.1.

In 2020, the share of the poorest 50% in total inheritance was 6% in Europe (Britain–France–Sweden average), with 39% for the middle 40% and 55% for the richest 10%. After implementation of a minimum inheritance equal to 60% of average wealth, financed by a progressive tax on both

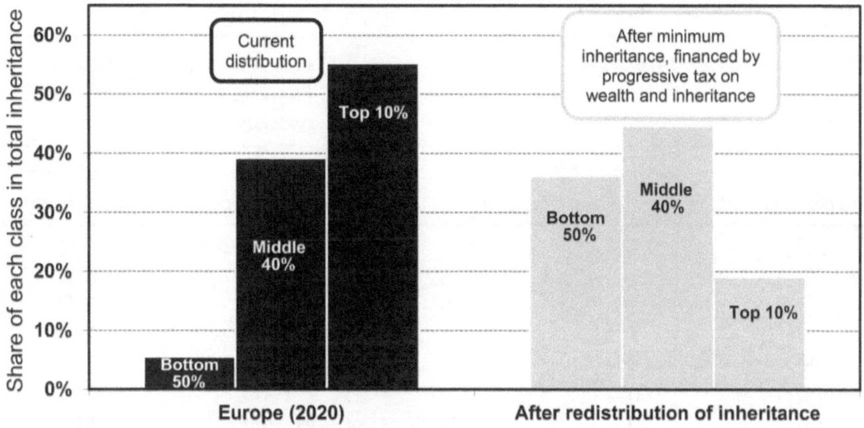

FIGURE 8.3.1 The distribution of inherited wealth, current, and projected.

Source: Adapted from Piketty (2022) [8A].

wealth and inheritance, the shares would be 36%, 45%, and 19% for the three groups. Given the 50:40:10 ratio of population in the three groups, these shares correspond to ratios of about 7:11:19 for the average individual in each group. As now, there would be wider disparity within the top 10% than in other groups.

The proposed tax system to finance this redistribution includes a progressive tax on property (an annual tax plus inheritance tax) funding the capital endowment for young adults, together with a progressive (roughly logarithmic) tax on income to fund both the basic income and public services – health, education, pensions, unemployment, energy, and so on.

In the example shown in Table 8.3.2, the progressive property tax amounts to about 5% of national income – enough to fund a capital endowment of about 60% of average net wealth to be allocated to each young adult at 25 years of age. The progressive income tax raises about 45% of national income – enough to fund an annual basic income of about 60% of average after-tax income, costing about 5% of national income, and the public services of a social and ecological state costing about 40% of national income.

ACTIVITY

In the light of the data presented in this section, discuss the pros and cons of the Piketty model from both 'fairness' and 'economic' points of view, as well as 'levers' identified by Wilkinson and Pickett.

TABLE 8.3.2 Piketty's model for circulation of property and progressive taxation

Progressive tax on property *Funding capital endowment for each young adult*			*Progressive tax on income* *Funding basic income and services*	
Multiple of average wealth	*Annual tax*	*Inheritance tax*	*Multiple of average income*	*Effective tax rate*
0.5	0.1%	5%	0.5	5%
2	1%	10%	2	10%
5	2%	50%	5	50%
10	5%	60%	10	60%
100	10%	70%	100	70%
1000	60%	80%	1000	80%
10 000	90%	90%	10 000	90%

Source: Data from Piketty (2022) [8A].

While this model may seem extreme, and the inheritance aspects are unprecedented, the top income tax rates are comparable to those operating in the United Kingdom and the United States in much of the 1950s when growth was faster and inequality was lower than it has developed since 1990 (Figures 8.1.3 and 8.1.4 show the US data). The top rate of income tax in the UK was around 90% in the 1950s but has dropped to about 45%. Inequality in the UK increased sharply during the 1980s and is now among the greatest among the OECD countries (see Figure 8.2.1). Understanding issues like these is an important goal of mathematical literacy for informed citizenship. In a democracy, there will always be different opinions on the best way forward, but making good decisions depends on knowing the broad parameters under which your society is operating.

References

Piketty, T. (2022). *A brief history of equality*. Harvard.
Wilkinson, R., & Pickett, K. (2009). *The spirit level*. Penguin.

Acknowledgements

Figure 8.2.1 is based on a figure taken from Wilkinson and Pickett (2009); the original is available in a slideshow available via link [8B] below. All of the other figures are adapted from Piketty (2022) and appear courtesy of the World Inequality Lab. The originals can be found at link [8A] below.

Links to useful material

To visit any of these links, scan this QR code or visit ltml.mathlit.org – append the link code to go directly to the entry – for example, ltml.mathlit.org/8A

The original source links are given below for attribution purposes:

[8A] *A Brief History of Equality* – Thomas Piketty
http://piketty.pse.ens.fr/fr/equality

[8B] *The Spirit Level* – Equality Trust
https://equalitytrust.org.uk/resources/the-spirit-level

9

YOUR MONEY IN YOUR LIFE

When thinking about using mathematics in everyday life, money is the first domain that most people think of. Control of your finances has always been an important ingredient in human happiness. It is also worth being more sophisticated than Mr. Micawber's advice to David Copperfield, in the context of Victorian England:

> Annual income 20 pounds, annual expenditure 19 pounds, nineteen and six, result happiness. Annual income 20 pounds, annual expenditure 20 pounds ought and six, result misery.
>
> Charles Dickens (1850). *David Copperfield, Chapter 12*

The need to "reconcile my net income with my gross habits" (attributed to '30s film star Errol Flynn) encapsulates an important strategy for everyone. But it is not specific enough to be much help on its own. What are the issues? What questions should we ask ourselves in the modern world?

This chapter aims to bring out underlying principles that enable people to get a clear view of where they are financially, and the options that are available to them on different timescales – but without their spending more time on details than their circumstances and inclinations demand. It is written on the assumption that those involved will live and work in a developed country and will be empowered by a deeper understanding of issues such as debit, credit, investments, consumer choices, and other aspects of personal financial management. While the principles that are discussed are general, some of the specific situations we discuss will be relevant only to some people, and only in the future. As 'believable' rather than 'action' problems, they are an important part of mathematical literacy. Indeed, the issues discussed are mainly the concern of adults, so the examples used may be rather different to those that are relevant for young teenagers.

DOI: 10.4324/9781003303503-10

We do *not* aim to provide a course on financial literacy, which in recent years has become an important topic in schools with many curriculum materials developed (see, for example, Sawatzki, 2016). Financial literacy has been assessed in every PISA cycle from 2012, so basic data on the financial knowledge of 15 year olds around the world is accumulating. The framework for the 2022 PISA financial assessment [9G] provides an overview of findings, and describes how the financial services directed to young people have changed. More services are now available, even to children as young as 5 in some countries, and new products are often aimed at young people, including 'buy now, pay later' purchasing with high penalties for missed payments.

Financial literacy courses are also available online and there is an ever-expanding range of apps that enable people to keep track both of their financial transactions and of the financial environment in which they operate. However, using the rich resources they offer tends to take a lot of time – itself a precious commodity. One definition of being rich is 'shorter of time than money'.

In getting to grips with the essentials of finance, as always with problem-solving, identifying the important variables is the starting point, moving on to explore the relationships between them. Here the key variables include income, expenditure of various kinds, capital, debt, growth of various kinds, and inflation; others will emerge in the course of the chapter. Overarching variables, and associated questions, include the following:

- *Timescale:* How are things financially at the moment? How urgent is it for you to get more money? How might you do it? Would you be willing to wait and study for a while, or work for lower pay for the benefit of experience, so as to get more in the end? If so, is waiting a few years OK?
- *Risk:* In financial matters do you accept, even enjoy the risk element, or, conversely, are you 'risk-averse'? Financial decisions, particularly long-term strategic ones, have some uncertainty in outcome – hence the standard phrase accompanying financial advice "may go down as well as up. Past results are no guarantee of future performance".

Questions like these are the basis of what follows – and of the learning activities we suggest. Money can be a sensitive subject, with people at very different levels of wealth, income, or debt in the same class. The activities here are structured to avoid direct comparisons that may be uncomfortable for some students. How they are used is, as always, a matter for the teacher.

Most issues in this chapter will be familiar to most teacher educators, many of them to their students, fewer to the school students they will teach who will have other experiences and perspectives. So the challenge is to find the most effective teaching approach to develop robust learning at each level. Discussion of the processes of financial literacy is central; on the content the challenge is to find an appropriate balance between a quick review (as here) and careful direct explanation of all the details. As usual, the suggested activities are structured to promote discussions that elicit the key issues, guiding the decisions that can help improve quality of life in the broad sense.

9.1 Knowing where you are

This is about strategy on the shortest timescale – and the least ambiguous issue, though not entirely without complexities. It offers a simple example of working at different levels of detail. Channelling Mr. Micawber, these questions will usually stimulate a lively discussion with a range of responses:

- Are we spending more than our income?
- Are we spending it on the things we really want? How can we check on this?

A detailed approach

A commonly recommended approach is detailed budgeting – keep a note of everything you spend in a week (or a month) and compare it to what you get in. There are many ways of doing this: from a simple notebook to a spreadsheet or an app designed for the purpose. The process is somewhat laborious, but the detailed picture it provides is crucial to gaining mastery over one's finances. To be really useful, this approach needs expenditure to be broken down under various categories.

Figure 9.1.1 shows a simple spreadsheet layout; the formulae involved can be seen by downloading the spreadsheet [9A].

Income and Expenditure

Start date:	01-Apr-2022	**End Date:**	30-Apr-2022	**Interval:** 4	**Weeks**

Income

Date	Source	Amount
17-Apr-2022	Avalon Lotteries	$ 10.00
26-Apr-2022	Wages	$1,150.00
	Total:	$1,160.00

Summary

Income:	$ 1,160.00
Expenditure:	$ 1,134.10
Cash Flow:	$ 25.90

Expenditure

Date	Description	Housing	Utilities	Goods	Groceries	Travel	Entertainment	Other
01-Apr-2022	Lottery ticket						$ 2.00	
08-Apr-2022	Weekly shopping				$ 43.44			
10-Apr-2022	TunezStream						$ 9.99	
12-Apr-2022	Bread and milk				$ 10.75			
15-Apr-2022	Weekly shopping				$ 55.37			
15-Apr-2022	New shoes			$124.99				
16-Apr-2022	Medicine							$ 10.59
17-Apr-2022	Fuel for car					$43.29		
19-Apr-2022	Car servicing					$50.00		
22-Apr-2022	Broadband bill		$ 37.95					
22-Apr-2022	Weekly shopping				$ 48.22			
23-Apr-2022	Plant food			$ 6.95				
26-Apr-2022	Bread and milk				$ 11.12			
29-Apr-2022	Rent	$ 500.00						
29-Apr-2022	Electricity bill		$117.29					
30-Apr-2022	Weekly shopping				$ 39.20			
30-Apr-2022	Mobile phone bill		$ 22.95					
	Totals:	$ 500.00	$178.19	$131.94	$ 208.10	$93.29	$ 11.99	$ 10.59
							Total expenditure:	$ 1,134.10

FIGURE 9.1.1 A spreadsheet for detailed accounting.

Source: Spreadsheet available online [9A].

The detailed information, particularly the totals for each category, give a picture of where the money is going; this may be compared with an expenditure plan – the 'budget' – or may inform the process of making one.

ACTIVITY

Everyone in the class tries this for a week, or a month, choosing their own categories. In the subsequent discussion, the teacher might ask questions like these. How did the process work out for you? Did you carry it through? Did you learn anything useful about your finances? I don't want numbers; just your experiences of the process.

To repeat: it is clearly important to avoid surfacing actual numbers for income or expenditure – some members of the class will have much more money than others. The sample spreadsheet offers a number-free pie chart (Figure 9.1.2) showing the spending broken down into categories, which could be a less sensitive subject for discussions.

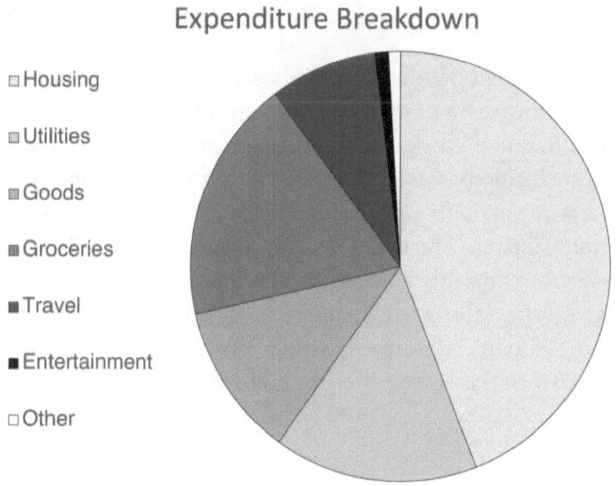

FIGURE 9.1.2 Number-free pie chart for comparison of expenditure.

Source: Spreadsheet available online [9A].

The big picture

This may be the point to present a question of strategy for discussion. Is there a way to cut through this complexity? Can you find a simple way of knowing where you are, month by month, without tracking the details? Ideas will surely emerge, including some of the following.

Historically, when all payments and earnings were in cash, you could add up all your money once a month (or once a week) and compare it to the previous month. This was a straightforward route to the big picture of your current finances. Adding a bank account meant the change in the balance over the month had to be included, as well as payments made in other ways such as cheque or lay-buy or hire purchase or debts.

Now people generally have many different accounts, but with technology you can check all your accounts online on the same day. The details are changing rapidly as technology evolves, but the basics of income versus expenditure and the cost of credit remain, and most online or mobile payment services ultimately link to a credit card or bank account. (The emerging, and controversial, world of cryptocurrency might see that disrupted in the future.)

One issue that may arise is how you treat items bought on a credit card. Conservatively, as expenditure on the day that each item was incurred, even though you won't pay them for up to six weeks – or much longer if you don't pay off the balance each month. How do you account for future expenditure incurred but not yet made, such as direct bank transfers or other periodic payments? That takes the conversation into issues such as debt and interest that go beyond 'knowing where you are'. We come to them later.

As things begin to look more complicated, keeping a complete record of all your expenditure may sound like a simplification but, as we saw, it involves a lot of detailed work – always something to avoid unless it is essential. There is a technology-based alternative that many people use – to make all your purchases through one account, for example, a debit card or by electronic funds transfer. The early assumption in the COVID-19 pandemic, that transmission was mainly through touching infected surfaces, pushed this change forward, especially with the advent of 'contactless' payments. When combined with the bank statement, which gives income and direct transfer payments, this can make seeing the big picture of month-to-month changes easier.

ACTIVITY

Compare detailed tracking with a big picture approach. Keeping a detailed record of expenditure is time-consuming. How often, and when, is it worth the effort?

In practice, many people who are living comfortably within their income don't track their expenditure in detail but keep an eye on 'the big picture' from time to time. However, when circumstances change – new place or new job, for example – it makes sense to monitor things more closely.

ACTIVITY

Discuss the challenges that arose in this discussion – mainly identifying variables and then finding, organising, and processing information. What have been the challenges in thinking through your current finances? After all, the mathematical concepts and techniques are simple – essentially just arithmetic.

This nicely illustrates the differences between strategic, tactical, and technical elements in the overall difficulty of a task – and how, in contrast to most school mathematics which is technically demanding for the student, the first two are often the main challenges in developing mathematical literacy.

Capital versus cash flow

This is a good point to surface the important distinction between these basic concepts. Capital is how much money you have at a point in time; the units are money: £, $, €, and so on. Cash flow, *income – expenditure*, is a rate – the clue is the word 'flow'; the units are money/time: £ per year, $ per month, for example.

Detailed record-keeping notes every 'in' or 'out' event over a period. The big picture approach calculates cash flow by measuring capital at two different times and dividing by the interval, avoiding having to calculate everything directly in detail.

ACTIVITY

In preparation for a discussion, ask each student to calculate their cash flow in a recent month by both methods, detailed versus big picture. Then they should compute the *percentage difference* between the two numbers – if any. Later, small groups discuss the advantages and disadvantages of the detailed budgeting versus big picture approach. They report on findings of differences, and evaluate accuracy against effort.

Credit versus debt

This important distinction is mainly one of attitude – in both cases someone owes money, often you! Credit is a particularly slippery term. Credit is often used for money we have received on the agreed understanding that we will repay the 'loan' later in a well-specified way. Your credit rating reflects a history of how well you met these promised repayments in the past. A good credit rating will be used as a guide in future by people considering giving you credit for something you want.

But on some occasions, you are the creditor. When your bank account is 'in credit', meaning a positive balance, you are lending money to the bank – on the understanding that you can recover it on agreed terms – immediately you want to or after a specific period.

Debt is a more general term for money you owe, and unless it is a loan from a friend or family, it is likely to incur regular interest payments at a specified rate. It may or may not have a specific date for 'paying off the debt'. Failure to make the payments will have consequences – financial, or sometimes more serious.

Ironically, credit cards offer free credit up to the monthly repayment date when they essentially become 'debt cards' – with a high rate of interest on the balance. In contrast, 'debit cards' draw directly from your bank account; they will only pay out if you have money in your account, so no interest is involved.

9.2 Where do you want to go? The issue of values

The obvious answer is "I would like as much money as possible" ("within the law" is sometimes added or implied), but this is rarely true – most people have limits to what they are prepared to do for money. It leads on to other questions. Financial literacy courses often present this issue as *Need* versus *Want*, but this too is simplistic. Answers to such questions are value-based and personal, often influenced by social pressures as well as circumstances.

What is a 'need'?

Let's start with the shortest timescale: current needs. Beyond the basics of food, shelter, and safety, many things over the last century have started as novelties but later come to be regarded as necessities: a car, a machine to wash and dry clothes, a dishwasher, a vacuum cleaner, a microwave oven – all these are practical aids to making life easier. While they are not, strictly speaking, essential to life, the pace of modern society – especially the expectation that mothers as well as fathers will have jobs – makes them hard to do without.

This concept of 'necessity' is in continuous flux: the 'landline' telephone, radio, and television started as novelties for the rich, became near-necessities but are now being replaced for some people by the smartphone with internet access. Social changes have made gym subscriptions, subscriptions to sports

TV/video channels, and fast fashion feel essential for many. The rise of the 'nail bar' is a vivid example of something moving over a few decades from invention to being a 'want' and then, for some, a 'need'. There are also activities that might be regarded as 'needs', such as playing sport, learning music, going to the cinema, travel, or going out with friends.

ACTIVITY

Introduce by asking: Beyond enough money to stay dry, warm, and fed, what are the needs on which most people agree? Then,

- ask each person to list their needs and their wants, and then to rank order items in each set;
- compare lists within a group, identify differences, and try to elicit underlying factors – and any principles that seem to emerge; and
- ask a spokesperson for each group to present their lists and answer questions.

How far down each rank-ordered list a person can go is determined by how much money they have (or can borrow – we will return to that). If 'getting as much money as possible' is your overriding goal, it implies that you stick to your 'needs' list – excluding everything else. The discussion will usually show that this is rare.

Thinking longer term

In thinking on a longer timescale, the financial scale is correspondingly larger. Currently, the order of magnitude of the lifetime earnings of a typical person who works for 40 years is around $1 000 000 after taxes, but with big variations depending on background, career choices, support, and luck. To take an obvious example, the decision on whether to go to university or to get a job straight from school is a complex trade-off between

- immediate income versus the prospect of earning more after a deferred start;
- higher salaries available to many graduates versus the long-term costs of paying off student loans;
- the benefits of the university experience; and
- a potentially more interesting job.

Estimates of increased lifetime earnings with a bachelor's degree – which some say is about 20% in real terms – are dependent on assumptions about the future of jobs and the economy that are uncertain, to say the least.

ACTIVITY

In small groups, explore the estimates of income for different jobs, bearing in mind how long or expensive the training is. Discuss trade-offs such as those listed above. Some students may be interested to find data to improve their estimates.

Values and goals

The questions a good financial adviser asks a new client will be designed to understand the client's values and priorities in some depth. The first step is to ask the client to discuss what they want to achieve, including things like security, freedom, job satisfaction, time for family and friends, a sense of accomplishment, making a difference to other people's lives, and work–life balance. The adviser uses this information to develop a sense of the client's priorities.

The next step is to turn these values into specific goals, with target dates, the money needed to reach them, and – less obvious, perhaps – the satisfaction that achieving each is likely to bring. Together, these form the basis of a 'financial road map' to guide planning (see, e.g., Bachrach, 2000). This is not a 'one off' activity but is reviewed regularly as circumstances or values change.

ACTIVITY

Discuss the value of a process that claims to measure client values and financial goals.

Setting financial goals in one's youth may well seem unrealistic but, in reality, this is just the right time – essentially because of the power of the exponential. It seems obvious that an individual with clear goals is likely to modify their behaviour over time, and very likely their financial habits, in pursuit of them.

9.3 Growth and decay

In the section on epidemics in Chapter 4, we saw the increasingly rapid growth that characterises exponential behaviour – and how difficult it is for us to grasp this behaviour intuitively. Financial thinking reflects this, in both directions. A modest amount of capital invested at an apparently low interest rate, compounding over many years, grows into something substantial. The power of compound interest to multiply capital over long periods is a concept at the core of good financial planning. Getting to understand exponential behaviour is core mathematics and central to many aspects of mathematical literacy.

Exponential growth and money

A spreadsheet like that in Figure 9.3.1 is a good way to see how the exponential grows, row by row, faster and faster. The graph shows how the growth is mostly towards the end, particularly over many 'doubling times'. The famous 'grains of rice on a chess board' problem, described in Chapter 2, is mirrored

	% per year*					
Interest rate:	**100**	**50**	**25**	**10**	**5**	**3**
Year	**Value**					
0	1	1	1	1	1	1
1	**2**	**1.50**	1.25	1.10	1.05	1.03
2	4	**2.25**	1.56	1.21	1.10	1.06
3	8	3.38	**1.95**	1.33	1.16	1.09
4	16	5.06	**2.44**	1.46	1.22	1.13
5	32	7.59	3.05	1.61	1.28	1.16
6	64	11.39	3.81	1.77	1.34	1.19
7	128	17.09	4.77	**1.95**	1.41	1.23
8	256	25.63	5.96	**2.14**	1.48	1.27
9	512	38.44	7.45	2.36	1.55	1.30
10	1,024	57.67	9.31	2.59	1.63	1.34
11	2,048	86.50	[illegible]	[illegible]	[illegible]	1.38
12	4,096	[illegible]	[illegible]	8.14	2.93	[illegible]
[illegible]	[illegible]	[illegible]	169.41	8.95	3.07	**1.97**
24	16,777,216	16,834.11	211.76	9.85	3.23	**2.03**
25	33,554,432	25,251.17	264.70	10.83	3.39	2.09
Doubling time:	**1.00**	**1.71**	**3.11**	**7.27**	**14.21**	**23.45**

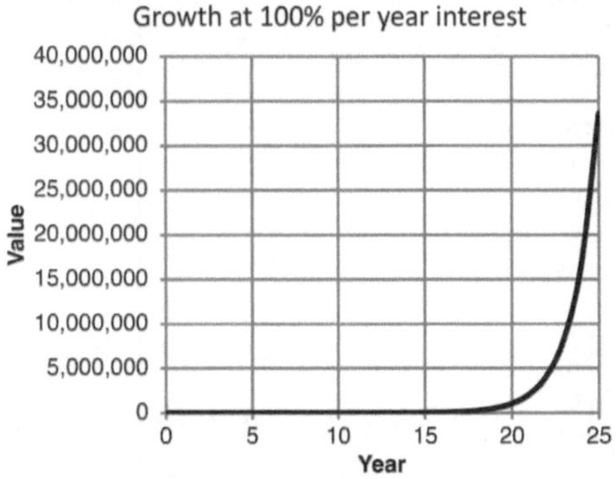

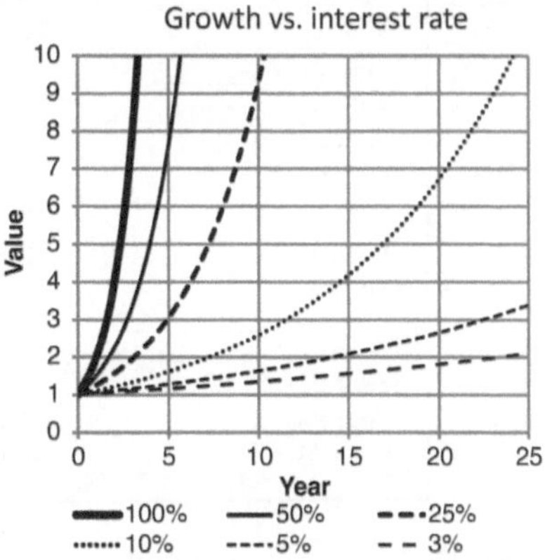

FIGURE 9.3.1 Doubling time investigation.

Source: Spreadsheet available online [9B].

in the '100% per interval' column in Figure 9.3.1. Other columns represent growth with various interest rates – which are usually much larger for debts than for savings! It may be worth spending time discussing the implications of how the result changes with the interest rate. Explore the concept of 'doubling time' through examples, as in the figure. The live spreadsheet used for the figure can be downloaded from link [9B].

ACTIVITY

Adam is 25. He is looking to save money for a pension starting with £100. He has been offered an investment paying 3% per year interest. This can be investigated with a simple spreadsheet (Figure 9.3.2 – link [9C])

- If he leaves the interest on his initial £100 to 'compound' in the investment, how much will it be worth at age 65? (£326)
- If he continues to add £100 each year, what will the total be? (around £8000)
- How does this compare with the total he has invested? (£4000, i.e. doubled)

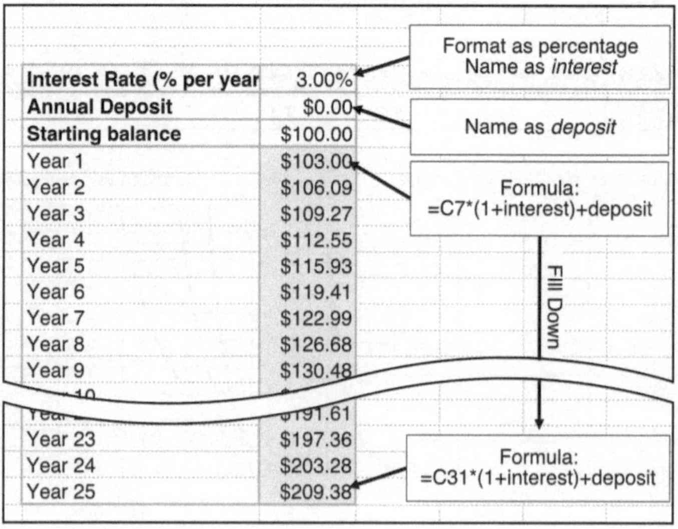

FIGURE 9.3.2 Simple savings calculator.

Source: Spreadsheet available online [9C].

More challenging mathematics

Some groups will be able to work out the doubling time in the general case by solving, for *n*:

$$\left(1+\frac{i}{100}\right)^n = 2 \text{ for } n, \text{ giving } n = \frac{\ln 2}{\ln\left(1+\frac{i}{100}\right)}$$

This solution is included at the bottom of each column in Figure 9.3.1.

There is a useful 'rule of thumb' which says that, for small interest rates, doubling time in years is approximately 70 divided by the annual percentage rate of interest. This is derived from the solution above by taking the first term from the Taylor's series expansion of the denominator:

$$\ln\left(1+\frac{i}{100}\right) = \frac{i}{100} + \frac{i}{100^2} + \cdots \approx \frac{i}{100}, \text{ so } n \approx \frac{\ln 2}{\left(\frac{i}{100}\right)} \approx \frac{100 \ln 2}{i} \approx \frac{70}{i}$$

(Note that this same formula was derived differently in Section 4.4 in finding the time for cases to double in an epidemic. The exponential function links both contexts.) To go into the mathematics more deeply, this situation provides an opportunity to relate the discrete mathematics of stepwise growth to that of continuous change (sometimes offered by US banks as "instant interest"). How do difference equations relate to the associated differential equation?

ACTIVITY

Investigate what interest rate per month/week/day/instantly corresponds to (say) 10% per year.

Inflation

Figure 9.3.3 shows UK prices since 1820 that are equivalent to £100 in 2020. Note that over many years in the nineteenth century, the prices of goods remained much the same but since World War I that has changed profoundly, with prices rising year-by-year more or less rapidly. Note the greater clarity of the logarithmic scale when things change by orders of magnitude, as here where about £1 in 1820 would buy as much as £100 about 200 years later. The issues involved in comparing prices over such a long period when buying patterns have changed are discussed in O'Donoghue et al. (2004).

Change in the value of money over time has led to the distinction between the cash value of something and its real value when allowance is made for

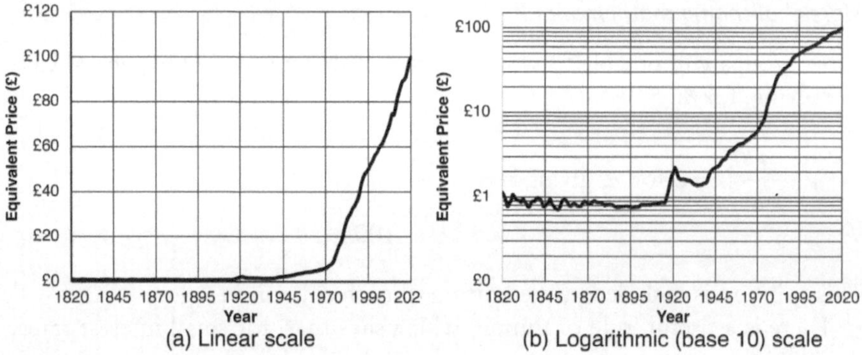

FIGURE 9.3.3 Equivalent prices in UK, 1820–2020.

Source: Data from O'Donohugh et al. (2004) and UK Office for National Statistics [9J].

inflation. So in the previous activity, if the inflation rate was 3% per year over the 40 years, the investment at 3% interest would be unchanged in *real value*. It has only *held its value*. In this artificial example, the growth from interest earned has been matched by the decay in real value from inflation.

On shorter timescales, inflation is often measured by comparing the price year-by-year of a 'basket' of goods that reflects what a typical consumer buys – it is essentially the gradient of the curve in Figure 9.3.3a; the percentage increase is the *consumer price index*, shown for the same data smoothed from year-to-year in Figure 9.3.4.

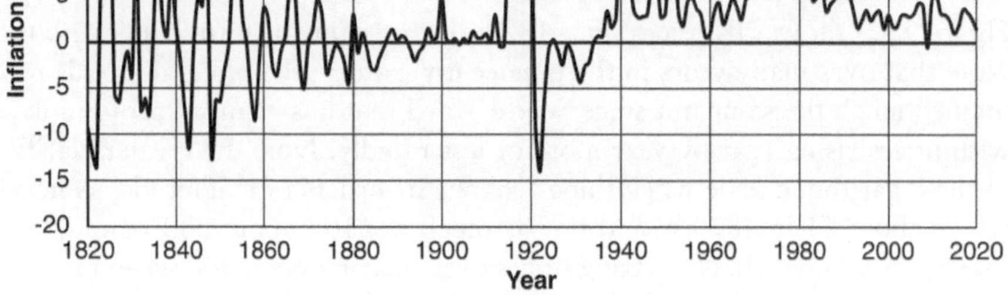

FIGURE 9.3.4 Consumer price index of inflation, 1820–2020.

Source: Data from O'Donohugh et al. (2004) and UK Office for National Statistics [9J].

Economists argue about the ideal rate of inflation – perhaps surprisingly, it is generally agreed that zero is not optimal. Why? Mainly to avoid coming close to deflation where prices fall. This has negative consequences for economic activity because it increases the real value of debt and discourages spending, because consumers expect prices to keep falling.

The growth of debt

Exponential growth applies equally to debt – and generally at a much higher rate than investment can return. The interest rates charged on credit card balances after the monthly payment date vary widely, but typical rates can be as high as 25% per year or even more. At this rate the spreadsheet in Figure 9.3.1 shows that without any repayment, the money owed doubles in about 3 years. Even the approximation $\frac{70}{i} = \frac{70}{25} = 2.8$ created for 'small' interest rates gives a salutary idea of the doubling time. Credit cards are convenient but almost any other type of loan, including a bank overdraft, is less expensive. 'Payday loan' rates are the exception; they are much higher again.

> **ACTIVITY**
>
> A discussion of debt and types of loan with pairs modelling explicit examples with numbers on a spreadsheet like that in Figure 9.3.1 will reinforce the potential value of slow thinking when it comes to incurring debt.

Nonetheless, debt can be advantageous in some circumstances – we then call it credit! Sellers sometimes offer very low, even zero, interest rates in order to clinch a sale, effectively spreading the cost over a period – with car sales often up to three years. However, this money is not free for the seller, so it must be built into the asking price.

9.4 Thinking longer term

Borrowing

There are circumstances in which borrowing capital can save a lot of money, despite Polonius' splendidly pompous advice in 'Hamlet':

> Neither a borrower nor a lender be
> For loan oft loseth both itself and friend
> While borrowing dulls the edge of husbandry.

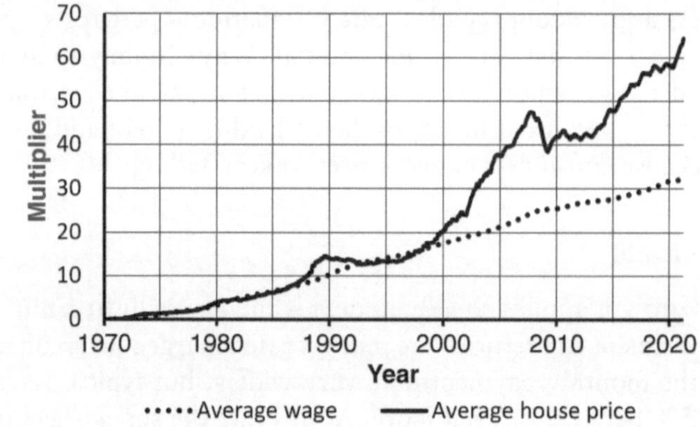

FIGURE 9.4.1 Change in UK house prices and average wages since 1970.
Source: Data from UK Office for National Statistics [9J].

Buying a house is the most familiar example. Houses in the UK have been an excellent investment over many decades, with prices rising much faster than average wages as shown in Figure 9.4.1 – and typically about 3% above inflation, although with periodic dips that have cost some buyers. Indeed, for most people with any assets, their house has been their main investment. So 'getting on the housing ladder' has rightly been a goal for many young people. However, for those with median incomes and no capital, it has proved impossible in recent years. Rents follow the upward trend, further 'squeezing' the standard of living of those who rent.

ACTIVITY

Discuss how rising house prices divide society into 'haves' and 'have nots'? What might be done to change the situation?

The graph in Figure 9.4.1 reflects other factors, including a link between house price increases and interest rates. In the early 1990s, rates touched 15% reducing demand for houses and thus prices, while the very low levels of interest rates in the 2010s have been a spur to house price rises. It may therefore be a mistake to assume that house prices will continue to generate such returns in the future, though the primary cause – the excess of demand over supply caused by the lack of sufficient housing – is not going to disappear any time soon. This is an interesting example of the many factors that can influence the price of a capital asset –a topic on which students may wish to study further.

Investment

It is unlikely that many students in a class will have significant sums for a *portfolio* of investments but they may be interested in role-playing financial planning scenarios. There is plenty of data on the performance of various types of investment over time. This can provide valuable experience in handling data – particularly in graphical form.

The fundamental choice is risk versus reward. Financial advisers complement the values issue with another set of questions to focus on how far the client wants to take risks in order to increase probable-but-uncertain returns. They look for consistent, at least semi-quantitative, measures of their client's appetite for risk. To this end, they have devised structured sequences of questions. Some are general and other questions present specific situations where the client chooses an option, as in the following illustrative examples:

- If you had to choose between more job security with a small pay increase and less job security with a big pay increase, which would you pick?
- Investments can go up or down in value and experts often say you should be prepared to weather a downturn. By how much could the total value of all your investments go down before you would begin to feel uncomfortable?
- Imagine that you are borrowing a large sum of money. It's not clear which way interest rates are going to move – up or down. Would you borrow at a variable interest rate or a fixed interest rate that is 1% more than the current variable rate?

From the answers to the questions, a client's risk score is calculated. This is another example of a mathematical model of an abstract concept (risk appetite) being created and used as a measure from which decisions can be made. Chapter 12 discusses the topic of creating measures. As with so many instruments used in social sciences, the operational issue is not whether the results are 'true' but whether they are consistent – which is evaluated during the development of the instrument – and whether they are useful for their declared purpose. Here, on the basis of the risk score, supported by the responses to individual questions, an adviser will recommend different mixtures of investment offering an appropriate balance of risks versus potential returns. In general, savings accounts with banks have low risk and low returns, bonds and unit trusts have more risk and moderate returns, while individual shares promise high returns at the cost of higher risk. Within each category, some products will have more risk and potentially more gain than others; for example, for shares in different companies, it depends on the kind of business as well, of course, as unexpected factors like management changes.

ACTIVITY

This 'investment game' gives students experience – direct but hazard-free – of the principles set out above. The app can be run via link [9E] and a screenshot is given in Figure 9.4.2.

The table and graph show the value of ten company shares over five years. Click on the table to see information on each company. This data is fictitious, but loosely based on actual stock market data for the period.

Work in pairs. If you are in a class, *don't* give away which shares you choose but *do* tell everybody how much money you have made or lost!

- Imagine you have £10 000 to invest and you can choose one company's shares to buy in 2017.
- Click on your company and click on the *Invest £10 000* button. The app will show you what happens to your investment.
- Share results with other pairs and discuss what you have learned about making and losing money.
- Now reinvest by dividing the £10 000 equally across five of the shares (without knowing how they would perform). Share and discuss the results and relate them to decisions about risk.

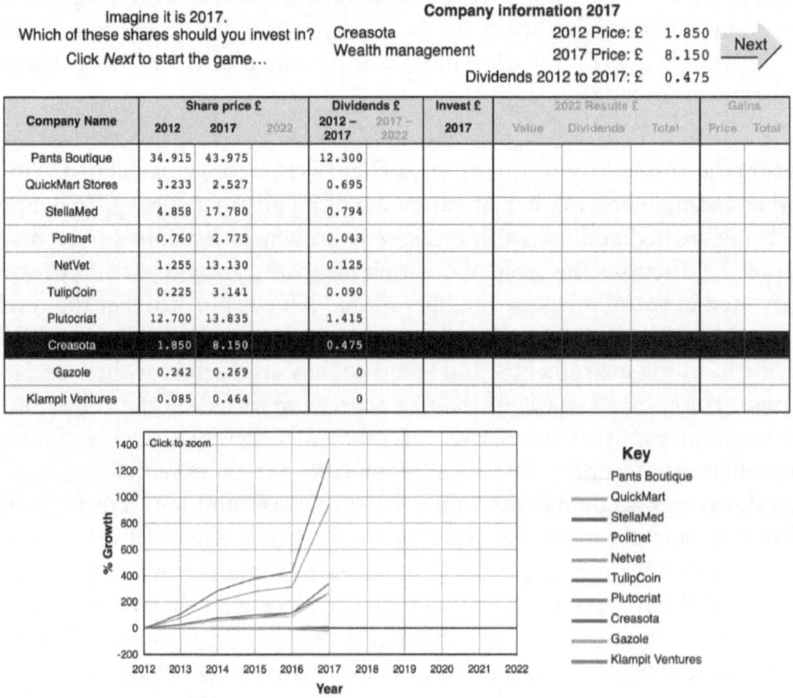

Company Name	Share price £			Dividends £		Invest £	2022 Results £			Gains	
	2012	2017	2022	2012 – 2017	2017 – 2022	2017	Value	Dividends	Total	Price	Total
Pants Boutique	34.915	43.975		12.300							
QuickMart Stores	3.233	2.527		0.695							
StellaMed	4.858	17.780		0.794							
Politnet	0.760	2.775		0.043							
NetVet	1.255	13.130		0.125							
TulipCoin	0.225	3.141		0.090							
Plutocriat	12.700	13.835		1.415							
Creasota	1.850	8.150		0.475							
Gazole	0.242	0.269		0							
Klampit Ventures	0.085	0.464		0							

FIGURE 9.4.2 The Investment Game. Can be run online [9E].

This already complex picture is actually greatly simplified. There are many key attributes of companies that a good fund manager will look for; for example, a strong market position, unique/protected products and services, and strong balance sheets. The finance industry has created, and continues to create, an enormous variety of 'products'. It makes money using your money. Finding your own way through all the possibilities is complex; getting professional guidance is normally wise.

9.5 Other topics

There are far more topics in finance relating to mathematical literacy than we can do justice to in this chapter. Some of the major omissions are outlined briefly below.

Basics

We have not attempted to cover important basic facts that are set out in multiple sources, and courses on financial literacy. For example, students need to learn to read documents such as bank statements, payslips, invoices, receipts, and tax documents, and there are many terms that need to be understood precisely.

Insurance

Here you pay a company with (you hope) plenty of capital to recompense you in the event of a specific loss or unexpected debt. As the company must cover its administrative costs and make a profit, a useful guiding principle is as follows:

> Don't insure against a loss that you can cover yourself without disrupting your lifestyle or plans.

So things like insuring for the maintenance of your domestic appliances after the manufacturer's guarantee ends – a common type of solicitation – is probably not a good deal. Indeed, this is the period of lowest risk of breakdown – after the initial 'bugs' have shown up and before wear and tear causes problems. There are many examples like that. The bathtub curve discussed in Chapter 7 (Figure 7.2.2) is relevant here.

There are types of risk that should be insured. Some are legally required – notably car insurance against damage or injury to someone else, the 'third party'. Some relate to other risks that you cannot afford to cover – notably loss or substantial damage to any house that you own, or the Stradivarius violin you are lucky enough to play and, in some countries, medical insurance. The language of insurance can be complex, including terms such as coverage, premiums, excess, exclusions, and the precious 'no claims bonus', which need to be well understood.

The present value of future income

In finance, 'discounting' is the process of determining the current value of a payment that is to be received in the future. This is an aspect of financial reasoning that is rather more subtle conceptually – and interesting mathematically. It reflects the fact that the money, if received now, could be earning compounding interest over the period of delay, though the *discount rate* is influenced by other market factors than the current interest rates – such as inflation.

The discounting factor D_n after n years at a constant discount rate of $r\%$ is given by

$$D_n = \frac{1}{(1+r)^n}$$

This is illustrated in Figure 9.5.1.

Further study

In this chapter we have focused on the big picture, on the fundamental principles that underlie financial literacy, and how they may be explored in classrooms. There are a number of professional bodies involved in the field of financial education, and some of these produce helpful financial guides that provide an introduction to most of the topics discussed above [9F]. An example is the Chartered Institute for Securities and Investment.

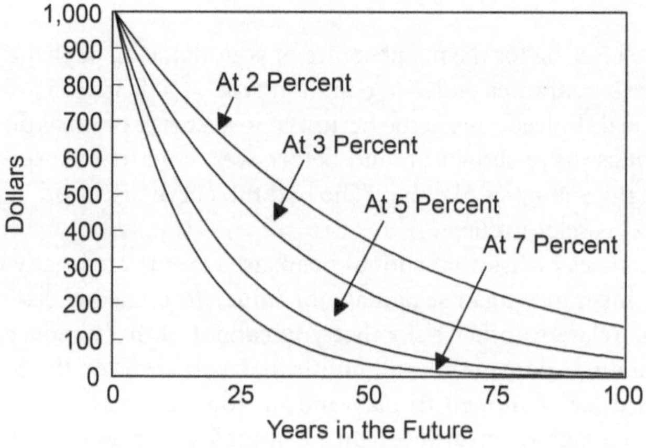

FIGURE 9.5.1 The present value of $1000 in the future at constant discount rates.
Source: Economics of Climate Change: A Primer – Ch. 3 [9H].

Acknowledgements

Our thanks to Andy Jervis and Joe Fawcett of Chesterton House Financial Services for their input on this chapter, particularly on investments. If you need real investment advice, please consult an expert directly!

Figures 9.3.3, 9.3.4, and 9.4.1 include data from the UK Office for National Statistics [9J], which is public sector information licensed under the Open Government Licence v3.0. The inflation data before 1949 in these come from O'Donohughe et al. (2004) – however, Kate Rose Morley's Historical Inflation Rates and Price Conversion Calculator at iamkate.com [9D] was helpful when preparing the section on inflation.

Figure 9.5.1 is taken from *Economics of Climate Change: A Primer*, public domain material from the US Congressional Budget Office (2003).

References

Bachrach, B. (2000). *Values-based financial planning: The art of creating an inspiring financial strategy*. Aim High Pub.

Dickens, C. (1850). *David Copperfield*. Project Gutenberg. https://www.gutenberg.org/ebooks/766

O'Donoghue, J., Golding, L., & Allen, G. (2004). *Consumer price inflation since 1750*. Office for National Statistics.

Sawatzki, C. (2016). Lessons in financial literacy task design: Authentic, imaginable, useful. *Mathematics Education Research Journal*, *29*(1), 1–19. https://doi.org/10.1007/s13394-016-0184-0

Links to useful material

To visit any of these links, scan this QR code or visit ltml.mathlit.org – append the link code to go directly to the entry – for example, ltml.mathlit.org/9A

The original source links are given below for attribution purposes:

[9A] *Knowing Where You Are* spreadsheet
https://ltml.mathlit.org/9A

[9B] *Doubling Time* spreadsheet
https://ltml.mathlit.org/9B

[9C] *Savings Calculator* spreadsheet
https://ltml.mathlit.org/9C

[9D] *Historical UK Inflation Rates and Price Conversion Calculator* – Kate Rose Morley
https://iamkate.com/data/uk-inflation/

[9E] *The Investment Game* app
https://ltml.mathlit.org/9E

[9F] *Sources of Financial Guides*
Chartered Institute for Securities and Investment https://www.cisi.org/

[9G] *Financial Literacy Analytical and Assessment Framework* – PISA/OECD
https://www.oecd.org/pisa/publications/pisa-2021-assessment-and-analytical-framework.htm

[9H] *The Economics of Climate Change: A Primer* – Congressional Budget Office
https://www.cbo.gov/publication/14387

[9J] *Inflation and Price Indices* – UK Office for National Statistics
https://www.ons.gov.uk/economy/inflationandpriceindices

10

COMPUTERS IN TEACHING FOR MATHEMATICAL LITERACY

10.1 Overview

There are many ways in which information technology (IT) can assist the teaching of mathematical literacy by putting powerful mathematical tools in students' hands, making real-world information easily available in the classroom, and presenting it in rich, interactive forms. Sometimes, this will also teach useful lessons about the use of such technology in work and personal life – but the time available, and the rapidly moving target of current technology, makes the teaching of valid, transferrable workplace IT skills a challenge, while the actual software tools used in the workplace can be highly specialised and complex to learn. So here we start by concentrating on ways IT can be used to help address the particular challenges of teaching for mathematical literacy, expanding on the multiple examples that appear elsewhere in the book. In Section 10.2, we start by looking at ways a computer can help in the modelling process, moving in Section 10.3 to roles in visualisations and simulations, and focusing on the use of video as a stimulus in Section 10.4. Section 10.5 addresses using realistic data sets, so central to mathematical literacy, while Section 10.6 is about error detection. Section 10.7 seeks to bring out the importance of units and orders of magnitude in technology. We conclude by pointing to some further opportunities in Section 10.8.

Of course, this chapter can only scratch the surface of the vast set of possibilities for using IT in teaching that are opening up to schools, and we have not attempted to address the teaching of coding, computer science, or the advanced use of software tools.

DOI: 10.4324/9781003303503-11

IT and the need for mathematical skills

Arguably, other subject areas have done more to embrace the use of technology – such as word processing, making presentations, researching information online, creating music or art – while, in mathematics, even 1970s pocket calculator technology is still sometimes contentious. Perhaps this is precisely *because* digital technology is so intertwined with mathematics – and thus disruptive to the traditional curriculum. A word processor won't write a *good* essay for you (although the qualifier 'good' is becoming important – see below), but the right program can compute the correct answer to an extremely complicated mathematics problem. The fundamental usefulness of a word processor doesn't vanish if you disable the spelling and grammar checker (which are major aspects of traditional school curriculum). By contrast, the fundamental usefulness of a computer algebra system is to compute answers as required in traditional mathematics problems (including step-by-step working if you wish) and a spreadsheet that doesn't do calculations is not very useful.

As we have seen, technology has changed the aspects of mathematics that are important, largely obviating the need to perform laborious calculations, graphing, or algebraic manipulations in most life and work situations. Instead, it has made it important to understand the underlying mathematics and how it models the situation at hand. It has not removed the relevance of more advanced mathematics to anybody wanting to pursue the subject into higher education. With vast amounts of data on hand, choosing the tools to analyse and represent data is essential, as is correctly interpreting those results – the procedural details of how to perform those analyses, or construct those representations less so. Any claims that repeated practice of a procedure will automatically engender better understanding of the underlying mathematics are not supported by research on learning; rather the cognitive load of mastering the procedure can overwhelm any deeper understanding. There are better ways to understand place value notation and associativity than practising long multiplication and division. As stated in Hoyles et al. (2007) when reporting on studies of "Techno-Mathematical Literacies" needed in the modern workplace: "the major skills deficit for mathematics in workplaces is the understanding of systems, not the ability to calculate". Later in this chapter we examine just two cases – spotting unreasonable results and dealing with 'new' units and dimensionality – of mathematics which has been given new importance by the ubiquity of information technology.

At the time of writing, the teaching of languages, arts, humanities, and social sciences, where assessment revolves around the essay or portfolio of work, appears to be facing an analogous challenge from widely available 'artificial intelligence' software that can write convincing (if uninspired) prose and poems and create paintings on a given subject. Here too, being able to use such tools correctly, honestly, and productively, and being able to understand and improve on the results, is likely to become the important life skill, requiring a change of emphasis in teaching somewhat analogous to those being discussed here for mathematics.

IT as a teaching tool

It is now common for mathematics classrooms in wealthy countries to have, at least, an electronic whiteboard with an internet connection – providing access to a vast bank of resources, some especially designed for education. Often, these resources are used to deliver fairly conventional course materials – but they have huge potential for introducing realistic content, such as genuine data sets and images, current news, and video material into a lesson. For nearly a century, there have been excellent educational films and videos, but their use has tended to be passive, whereas modern technology allows the more interactive use of short video snippets, animations, and interactive simulations that can help students engage with quite complex concepts and problems.

In the 'Making and Selling a Magazine' example lesson below, students go beyond straight-line graphs and locate the maximum of what looks quadratic but they use only basic arithmetic. In Chapter 9 it is shown how students can explore exponential functions, again using simple arithmetic and iteration. Technology makes it less tedious to explore different cases and avoids getting rice all over your chessboard. In the past, equations that cannot be easily solved algebraically had to be avoided. Now nearly *any* equation can easily be approximately solved visually by graphing it – in multiple dimensions if necessary – and zooming in on a root (if there is one!). Indeed, in industry and science, solving equations by (sophisticated) 'trial and improvement' is much more prevalent than solving equations analytically.

10.2 Computers in the modelling process

Chapter 1 discussed the mathematical modelling process and how it is widely applicable to many real-world applications of mathematics and described the cyclical 'modelling process' shown in Figure 10.2.1. In a traditional pen-and-paper mathematics classroom, the technical skills involved in the 'solve' step (actually called 'compute' in some versions of the modelling cycle) are often the focus of the lesson at the expense of the other phases – and also the bottleneck that restricts the complexity of the model, what mathematics can be used, the size of the data set, and how many iterations of the modelling cycle can be completed. 'Solve' is expensive in terms of the required skills,

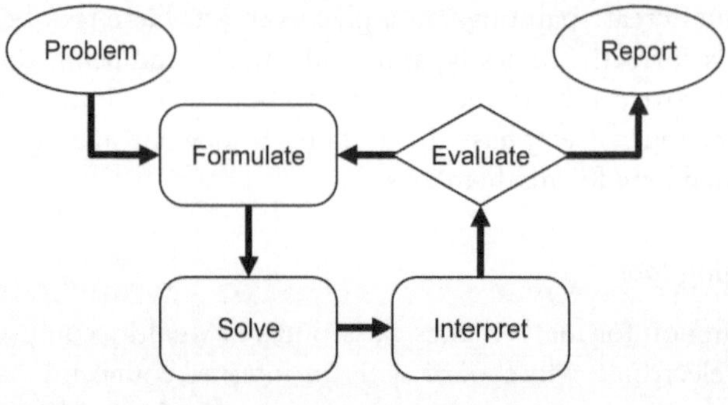

FIGURE 10.2.1 The modelling process. See Section 1.3.

the time taken in laborious calculations, and the total cognitive load on the students. With ubiquitous technology, 'compute' is often cheap, instantaneous, and can sensibly be a 'black box' with its internal workings remaining unrevealed in the lesson. Technology brings the opportunity to properly focus on the other three steps of the process.

Modelling with spreadsheets

Aside from the pocket calculator (whether in your pocket or on your wrist, phone, or computer), possibly the closest thing to a ubiquitous, general-purpose mathematics tool is the spreadsheet. Spreadsheets first appeared with *Visicalc* in the late 1970s and played a significant role in popularising early personal computers as serious business tools. Unlike older business software, spreadsheets were highly interactive. Their main original use was building and exploring "what-if?" financial models. The magazine task below is a very simple example. Over time spreadsheets have also become widely used as graphing, charting, statistics, and simple database tools.

In schools, the use of spreadsheets has tended to be dominated by handling survey data, particularly for collating in tables, summarising with simple statistics, and graphing. Less attention has been given to the use of cell references and formulae to build actual mathematical models. When the 'Making and Selling a Magazine' task in Figure 10.2.2 [10A] was being designed, it received criticism for being such a basic use of spreadsheets. However, during trialling, it emerged that while students *were* familiar with tabulating survey results, producing bar and pie charts and summary statistics, the idea of using formulae to turn that into a "what-if?" model was novel. The video even shows one pair using a calculator app to calculate values to type into spreadsheet cells.

The task in the Magazine lesson is to decide on the selling price for a product, covering the cost of producing the product and maximising profit.

4 Making and selling a magazine

Some teenagers want to raise money by making and selling a new magazine.
They conduct a survey to find out how the selling price might affect the number of people that will buy it.
They ask the following question to 100 people:

"How much would you be prepared to pay for this magazine?"

This is what they find:

Selling price (£)	0	0.50	1.00	1.50	2.00	2.50
Number of people who would buy it at this price.	100	82	60	42	18	0

Each magazine costs 10 pence to make.

What should the selling price be in order to make the most money?

Suppose production costs increase... how will this affect your answer?

Using a spreadsheet

Kim has started to solve this problem with a spreadsheet.
See if you can work out what she has done and use the spreadsheet to solve the problem.
(Start by changing Kim's name to your own!)

Kim's spreadsheet

Cost of making each magazine (in pence)	10	10	10	10	10	10
Selling price (£)	0.00	0.50	1.00	1.50	2.00	2.50
Number of people that will buy it	100	82	60	42	18	0
Money we get from selling this number (£)						
Cost of making this number (£)						
Total profit we make (£)						

FIGURE 10.2.2 Making and selling a magazine.

Source: Bowland Maths, 2008 [10A].

Will you make more money by selling lots at a low price, or fewer at a high price? The lesson design assumed that some students would only have minimal spreadsheet experience and so provided a template to ensure they could concentrate on the formulation aspect of the task and have time for discussion and reflection. A mathematically literate student should eventually be able to create this sort of spreadsheet themselves, using formulae, creating well-formatted graphs with sensible scales, and using a few more advanced tricks such as static references and named cells – all of which requires teaching time in mathematics or digital technology subjects. A video account [10B] of a lesson built around this task is available, including a useful example of how a teacher might introduce the scenario.

This task predates the decline of physical printed media, but it works equally well for the many other small products that students might make and sell (e.g. hair ties, decorative bag tags, snacks). It could start by asking the students to think of a product they could make and sell. The mathematics of supply, demand. and production costs remain the same.

The Magazine task is followed by a suggestion for a data gathering and analysis activity. Gathering their own data should help students engage with the subject. However, gathering data can be time-consuming, so it is important to allocate enough time for the later stages where students discuss the results, thinking critically about the model and its limitations, and the sampling. Experience of simple modelling like this can help build some 'know about' understanding of how modelling is used to answer serious problems.

Beyond using formulas and making graphs, a mathematically literate student should be able to use at least some of the most common built-in functions, such as sum and average and trendline. The 'Height and arm span' lesson outlined in Figure 10.2.3 is intended for students studying the algebra of linear functions in early secondary school. It begins by using the table and scatterplot functionalities of spreadsheets, but then uses the 'black box' capability to draw a trend line through data. At this early stage in learning algebra, the intention is to focus on the meaning of all four letters in $y = mx + c$, to link tabular, graphical, and symbolic representations and to see how functions might link to data and real-world problems. There is no intention for students to understand theory behind the trend line, for example, using least squares. Students seem to find the concept of a 'line of best fit' sufficiently intuitive. Many variations of this lesson are possible, matching different stages of students' learning. For example, instead of using the automated trend line, students could enter formulae for various lines, and use the spreadsheet to calculate a simple measure of goodness of fit by summing the absolute differences between observed and predicted heights. An alternative version, using a graphics calculator for easy graphing of test lines, is given by Asp et al. (2004, p. 8).

Height and Arm Span: A lesson on linear functions and data

In this lesson, the whole class investigates the claim that the arm span of a person is usually equal to their height.

Supply several tape measures (or possibly a measuring app on mobile phones) and have students work together to quickly measure and records each student's height and arm span (the maximum distance from fingertip to fingertip with outstretched arms).

Enter the data into two columns of a spreadsheet and share to group or individual computers or graphics calculators. Students create scatterplots and discuss whether the data supports the claim or not. Then the spreadsheet 'trend line' function is used to display a line of best fit (without formal definition), and the class can compare it with the claim ($y = x$). They discuss what the graph of the trend line shows, and how the trend line can be used for predicting arm span from height (or vice versa).

When the equation of the line is displayed, the class discusses the meaning of the numbers in the equation in mathematical and everyday terms. For example, if the equation is arm span = 1.2 × height – 6, the relationship can be described as "when a person grows by 10 cm, their arm span will probably increase by 12 cm". Discussion of the intercept will lead to observations about the range of validity of the trendline, and aspects of the data set, such as whether it is representative of the intended population. The automated option of setting the intercept to zero might be investigated. There are also opportunities to discuss other aspects of the data (e.g., outliers). The teacher could also supply some previously collected data from younger/older children or someone very tall.

FIGURE 10.2.3 Height and arm span: a lesson on linear functions and data.

Source: Adapted from Asp et al. (2004).

10.3 Visualisations and simulations

Technology has transformed the way data can be visualised. The summary statistics (mean, median, mode, correlation coefficient, etc.) typically studied in school mathematics are still important, but some of their usefulness harks back to the days when plotting a graph from a large data set was a time-consuming task. Looking at scatter graphs and other representations of data should now be a crucial first step in analysis. Whereas it is dangerous to draw conclusions from data without significance tests, it is just as dangerous to apply traditional line/curve fitting or summary statistics (all easy-to-use features of spreadsheets and statistical packages) without some idea of the distribution of the data. (Does a relationship look reasonably linear? Does a distribution look like it *has* a meaningful central tendency?)

Something as simple as a clear animation showing a process or a simulation based on entering one or two parameters can help to reinforce students' understanding of a concept. The 'Random variations' applet in Figure 4.2.4 in the section on 'Risk' is an example of a simple simulation used to illustrate a point. It shows the power of being able to run a simulation many times to see the effect of random elements. Using simulations is not just to reinforce mathematical concepts. With simulation as the 'third pillar of science' (Weinzierl, 2021) complementing the two long-standing pillars of theory and experiment, using simulations also helps to develop 'know about' insight into the way in which simulations can be used for solving complex problems.

There is some research (Ridgway et. al., 2007) to show that interactive visualisations on computers can allow students to engage with multivariate data more easily than when the information is presented on paper. The task shown in Figure 10.3.1 presented a simulated data set from a photosynthesis experiment in which the rate of oxygen production (dependent variable) was affected by two independent variables (temperature and light intensity). Students could put one independent variable on the horizontal axis of a graph and vary the other interactively using a slider. The actual task asks students to use the data from the graph to comment on a series of statements from fictitious students, such as "Jim suggests that they keep the plants as warm as possible; Amy suggests that they keep the lights on full power – who has the

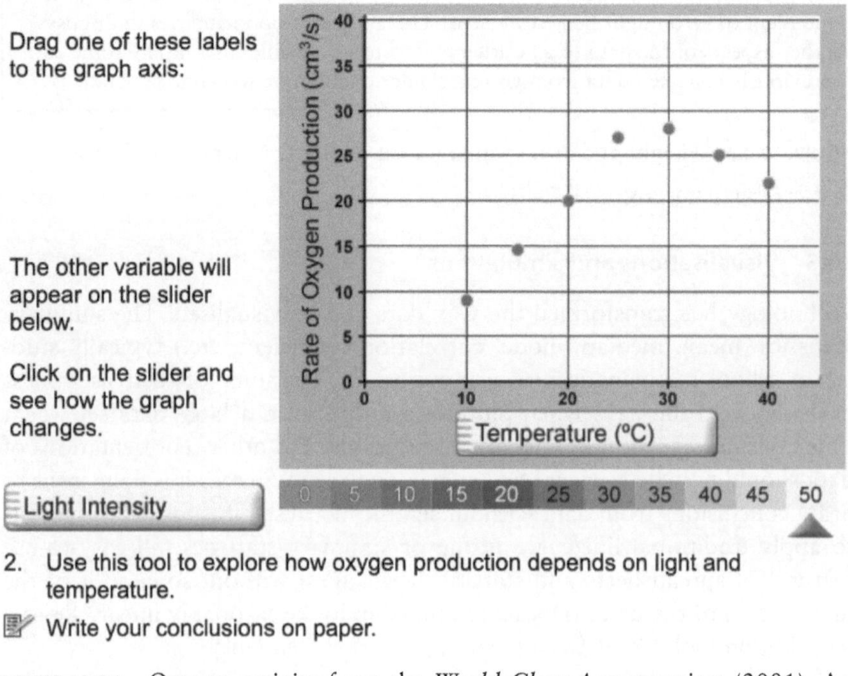

FIGURE 10.3.1 Oxygen activity from the *World Class Arena* project (2001). As discussed in Ridgway et al. (2007) [10C].

best idea?" In the study mentioned, performance on a series of tasks like this was compared with paper-based graph interpretation tasks which avoided the use of multivariate data. Despite the richer, more realistic data sets and the added complexity of multiple variables, students only found the multivariate tasks slightly more difficult than the paper ones.

Visualisation tools could be custom-written 'apps' or pre-prepared files for spreadsheets, graphers, or interactive geometry packages. In the reSolve 'Cornering' module, a series of lessons investigates the geometry of large vehicles turning corners, and the implications for the safety of other road users and for road design. To support this, the materials include a set of interactive geometry models of increasing complexity – supplied as files for the free interactive geometry tool GeoGebra – these help students visualise the situations and explore the effect of various parameters, such as the width of the vehicle and the length of the wheelbase. These interactive models help students visualise the situation, and as simulations of the real situation, they provide predictions about the space vehicles need to turn. All the online live models and files to download and perhaps modify are accessible from [10D].

The simplest situation is a bicycle or a scooter, and students can do their own experiments with them to get insight into the geometry of turning. Figure 10.3.2 illustrates the next stage, where students investigate the turning of a very long bicycle (a simplified long truck). Here the model primarily helps visualisation of a simplified case, for which students could reasonably

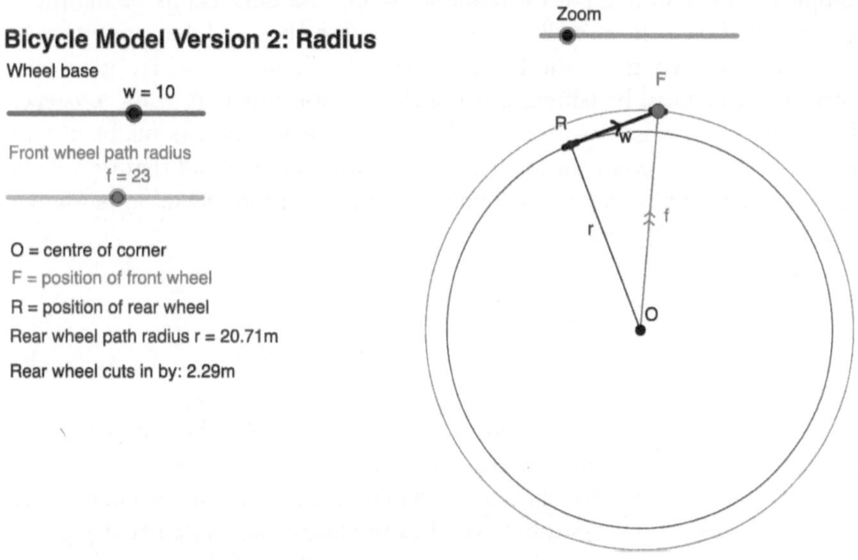

FIGURE 10.3.2 Bicycle cornering.

Source: From *reSolve: Maths by Inquiry (2018)* [10D].

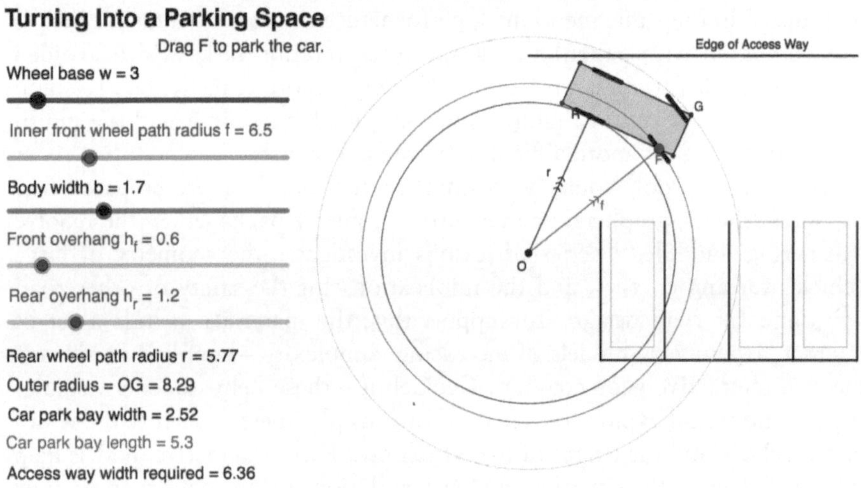

FIGURE 10.3.3 Turning into a parking space.

Source: From *reSolve: Maths by Inquiry (2018)* [10D].

be expected to follow the mathematics. With the interactive model, the emphasis can be on exploring the effects of the parameters, rather than repeatedly resolving the problem manually.

By the end of the unit, as in Figure 10.3.3, the model has become more complex, with multiple parameters controlling the dimensions of the (now four-wheeled) vehicle, as well as the size of parking bay and the turning space allowed. Although these models use freely available interactive geometry software, which can be edited, it is not the intention that students construct the model for themselves. A handful of enthusiastic students might like to do this, but at this point the students are drawing together all that they have learned, using the model to make and report the recommendations for the design of parking spaces.

10.4 Video as a stimulus

A short video or animation can be used to inspire students' interest and make the mathematics feel relevant. The use of "drama" in lesson design is discussed further in Chapter 11 and many of the examples there make use of stimulus videos. A straightforward mathematics activity using video stimuli is simply to present a video showing a changing situation, which students are asked to represent on a graph. Several examples of this are currently available on the Desmos website [10E] under the heading of 'Graphing Stories'. The lesson example "Will It Hit the Hoop?" is taken from the work of Dan Meyer and others: many others have created versions of this task. The video shows the start of several basketball shots, and students are asked to predict if each shot will go through the hoop and sketch the path. On the Desmos site, this has been built into an interactive activity with graph-sketching and curve-fitting tools for quadratic functions. Figure 10.4.1 shows a frame from the video at the top and the manipulable parabola at the bottom. Students

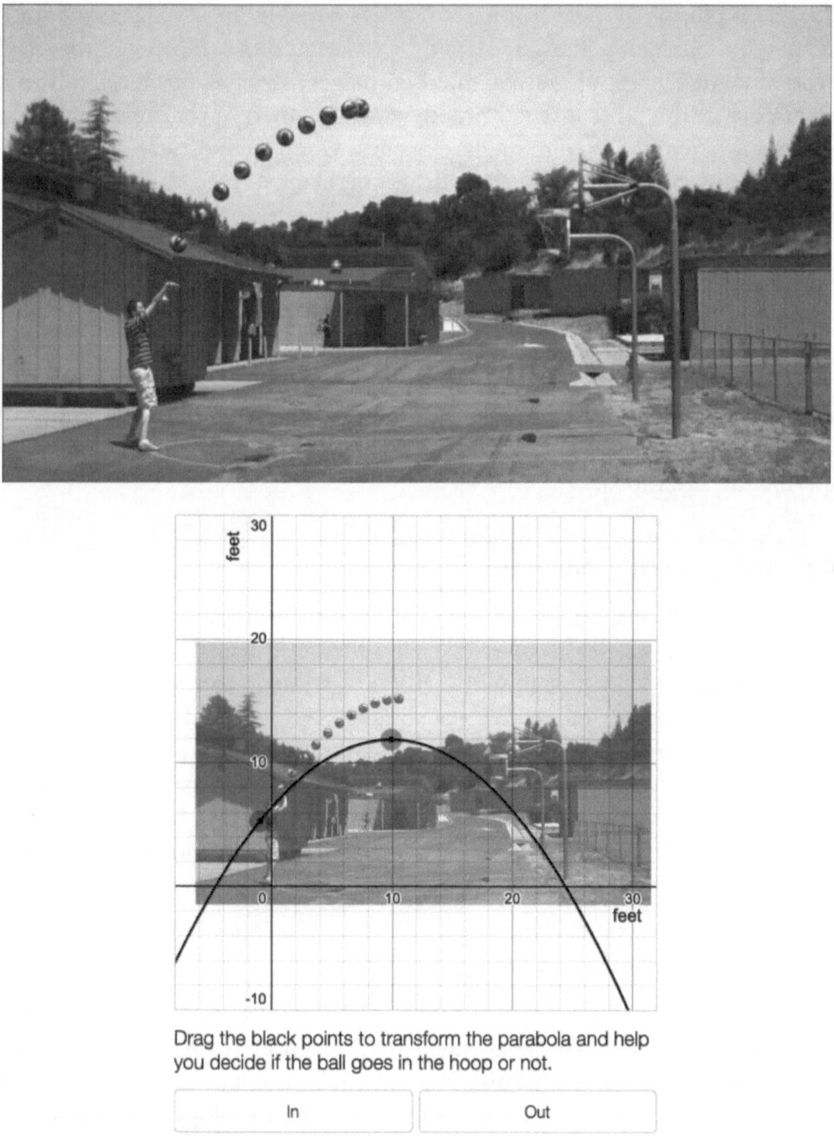

FIGURE 10.4.1 *Graphing Stories* example: video (top), curve fitting tool (bottom).
Source: From *Desmos* [10E].

use their best parabolas to make more predictions using the 'In" and 'Out' buttons supplied and then by seeing the full videos of the shots, compare the by-eye and parabola predictions. This is also an ideal way of practising plausible estimation so important to mathematical literacy; even the graphing activities generally require choosing a plausible scale.

Students will enjoy making their own videos for activities such as this or taking still photos of local features to analyse; maybe this bridge is a parabola, maybe a hanging chain or a rainbow is a parabola. However, they will need some advice about getting suitable images: straight on, level, without camera distortion. Spreadsheets and dynamic geometry programs (including Desmos' tool) can be used to analyse images. Good free software is available for analysing videos, and some students might have seen this when learning to play a sport or using sensor kits in science lessons. Pierce and Stacey (2006, 2011) provide many examples of using videos and images to learn about geometry, functions, and algebra.

Many of the example activities in this book – where they don't already use video or online data – could be enhanced by starting the lesson with, for example, a current clip from a news site. Videos and animations used in this way can help address one of the pitfalls of using realistic applications of mathematics in the classroom: the need for long, high-reading-age texts to describe the situations.

10.5 Using realistic data sets

For practical reasons, classroom data handling has traditionally been limited to tiny data sets which can be realistically managed using pencil and paper. Computers make it possible to work with larger, richer, and realistic data sets, and – as noted above – interactive visualisation and the freedom from the burden of calculation can allow students to explore more complex relationships that they would on paper. Many classrooms now have easy access to large quantities of real-world data online, including resources that have been specifically curated and developed for educational use. The following two lesson examples illustrate what can be done with good design and careful development.

Habitat modelling following fire

This lesson, intended for Year 9 and Year 10 school science and mathematics, is from the Victorian Department of Education *Numeracy Across the Curriculum* project [10F]. Ideally it is taught in conjunction with work in Science or Geography, so teachers with expertise in ecology can contribute. The lesson illustrates the mathematical literacy principle that the context needs to be taken seriously and gives a practical insight into how scientists use mathematical models. It draws on data collected from mallee habitat, which is widespread over drier parts of southern eastern Australia. Mallee is dominated by small multi-stemmed trees, sandy soil, and is home to many animals. Fire is a regular feature of mallee. The data used in this lesson was collected to study how the habitat responds in the decades after fire.

Scientists surveyed over 500 mallee sites with a known date of the last fire. They measured habitat features, like tree heights, litter, logs, and grasses and counted wildlife and plant species. The lesson is based on data that spans up to 110 years after fire for the variables related to the trees, *Triodia* (a dense spiky grass commonly called spinifex) and the 'legless lizard' *Delma australis*. Some of the data from the scientific publications is presented in a series of animations created by Ian Lund [10G]. The animation shows the changes over time. A column graph is used to show the value of each variable (tree height and cover, stem density, number of hollows, bark, numbers of *Triodia* and *Delma*). All the variables are scaled to a common range of 0–100. As the animation runs, the maximum value reached up to that time is marked with a red line on every column.

The lesson begins with teacher-led class discussion, ensuring that students understand the ecological variables and the data that are involved (e.g. what live hollows are and why they are important) and the way the animation presents the data over time. Students then might work in groups to look at the main trends shown by the animations. Some would be expected (trees get higher over time), whilst other trends are less easy to see and some are unexpected (tree cover reaches a maximum and then drops). Once students are comfortable analysing the seven habitat features over time, they work in pairs using a second animation which also shows the prevalence of the legless lizard. Students describe the prevalence of the lizard over the decades after a fire (it rises, slowly at first, then falls) and they identify the habitat features which are associated with this. Finally, students make predictions about what will happen in the short term and long term if there is a fire next summer, supporting their predictions with evidence and critiquing alternative conclusions from other pairs. Working together enables students to consider multiple interpretations, and become aware of other observations about the data.

This lesson illustrates a number of features about mathematical literacy. Understanding many aspects of the context is essential to interpret the data, and especially to make predictions from it. A teacher knowledgeable in ecology is needed to assist students learn why tracking changes in habitat over a long time after fire is important, what mallee habitat is like, why the selected variables are important, and how the data set has been assembled. The importance of work like this is heightened because as climate change is likely to cause more frequent burning and so the time between burns is likely to reduce – what will be the impact of this? At first glance, the only mathematics required is to be able to read column graphs, but many mathematical problem-solving skills and scientific inquiry skills need to be brought into play to deal with the complexity of the situation. Moreover, there are hidden mathematical aspects behind the presented data – how is data put onto the common 0–100 scale, how did the 500 sites provide data for a century, what sort of sampling was used to get the measurements (e.g. of the number of legless lizards), not all the trees will be of same height so what measure is used, and how reliable are the measurements of all variables likely to be. Central to this is the critical thinking that students have to engage in for testing out the potential patterns that they see, for providing evidence-based arguments to support their decisions, and for evaluating the arguments of others.

Open data example – literacy rates

Organisations such as the World Bank, UNESCO, Gapminder, and even the CIA publish large collections of data online that can be used to explore a wide range of topical social and economic issues – either through those organisations' own online visualisation tools or by downloading the raw data [10H]. There are also large scientific and environmental data sets, often with real-time updates available, many especially designed for education. The Australian science agency, CSIRO, for example, has educational data sets on topics such as the Great Barrier Reef, carbon dioxide concentrations, water quality in a dam, and inequality of income around Australia.

The World Bank data offers a wealth of demographic and economic data from around the world along with easy-to-use tools to graph it. The example below shows literacy rates over time. Data is also available showing the difference in literacy rates amongst males and females, as well as several statistics about schools and teachers that help to explore this topic. For example, Figure 10.5.1 shows two graphs of literacy of the whole world population created using the site. The 'gender parity index' is the female literacy rate divided by the male rate – how is that interpreted?

The site also allows the download of complete, detailed data sets that can be analysed using a spreadsheet or other tool. Figure 10.5.2 (overleaf) shows a "deeper dive" into the literacy data, with a scatter graph of male versus female literacy rates by country or region in 2021. The black line shows what you would expect if male and female rates were equal. Check that students know that if a point is on this line, it means the gender parity index is 1, and that nearly all the data points are below the line because female literacy is below male literacy nearly everywhere. Which are the exceptional countries, and what are their characteristics? What might be the relationship between male and female literacy? It is obvious that a straight line may not be the best fit. The dotted line shows the quadratic function:

$$f = \frac{m^2}{100}, \text{ where } f = \text{female literacy \%}, m = \text{male literacy \%}$$

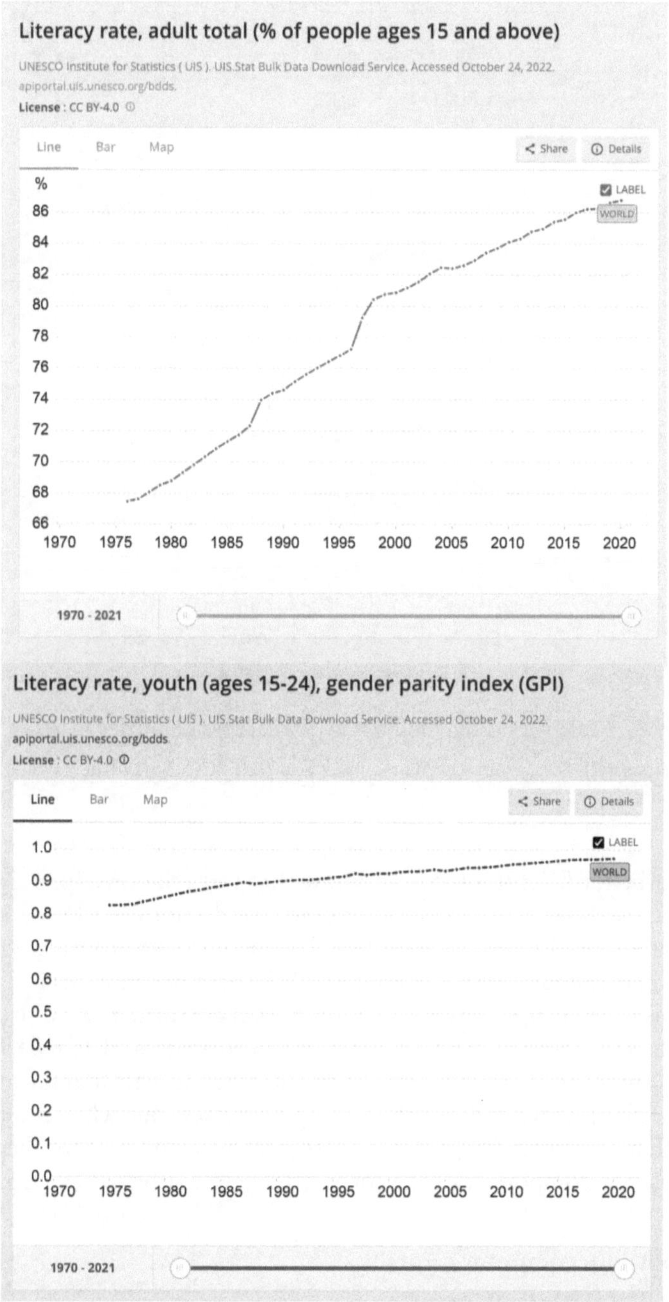

FIGURE 10.5.1 Graphs of the world literacy rate and gender parity index over time.

Source: Graph created on the World Bank Open Data site [10H].

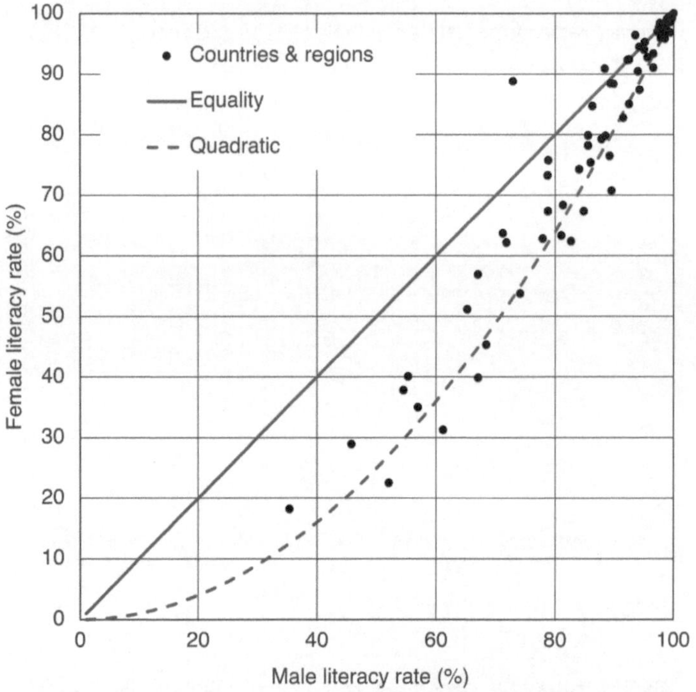

FIGURE 10.5.2 Male versus female literacy (%) across countries in 2021.

Source: Created with Data from The World Bank and UNESCO. Spreadsheet available online [10K].

which *is* a better fit. But we do not know that this relationship is significant in any way without more research; indeed, we do not even know if such a pattern would hold in other years. What is the purpose of fitting a curve to data like this? Is it just a useful summary of the data or can it be used to make predictions, or to propose hypotheses to investigate further? Does it give any clues as to the underlying relationship between male and female literacy? Certainly, this curved graph highlights the fact that as male literacy increases, the gap between male and female literacy decreases, but what could be the cause of that? (Chapter 11 discusses the role of curiosity – part of the productive disposition component of the context-focused mathematics framework.)

10.6 Spotting 'computer errors'

Here we move from using IT as a teaching tool to aspects of mathematics that take on new importance in life and work because of the rise of technology.

As the adage goes, "To err is human – but to really foul things up you need a computer!" Digital technology has the ability to repeat mistakes millions of times a second and computers have no common sense or intuition to spot when something has gone wrong. They follow the instructions supplied regardless. There are plenty of stories of careers and businesses being ended by an errant decimal point in a computer entry [10L], supporting the well-known saying in computing – 'garbage in, garbage out'. That this was not self-evident to most people was recognised by Charles Babbage (1791–1871) even before working computers existed:

> On two occasions I have been asked [by members of Parliament], 'Pray, Mr. Babbage, if you put into the machine wrong figures, will the right answers come out?' I am not able rightly to apprehend the kind of confusion of ideas that could provoke such a question.
>
> Charles Babbage (1864). *Passages from the Life of a Philosopher*, p. 67

So, given the possibilities of actual software bugs, user errors, or a combination of both, it is helpful to have *some* expectations as to what the correct answer might reasonably be, whenever using technology to perform a calculation.

Estimation skills have long been covered by most math curricula, often through estimation in the sense of *approximate calculation* rather than the broad sense required for mathematical literacy that is set out in Section 3.4. For example, students might be asked to estimate 199.8 ÷ 49.9, without detailed calculation. Often, as here, these questions are contrived to have only one reasonable answer, for the convenience of test scoring. This is not a bad problem, but it only covers one aspect of estimation and is arguably mostly a test of rounding skills. A more realistic "error spotting" test might be as follows:

Without using a calculator, decide which one of these calculations is wrong:

$199.8 \div 49.9 = 4.004$
$1980 \times 9.9 = 19800$
$1240 \times 0.1 = 124$
$1240 \div 0.1 = 12400$

A completely different – and usually more demanding – type of estimation task is the "plausible estimation" or "Fermi estimate" where students are tasked to estimate some quantity with very little of the information required for an accurate calculation.

In a video from Bowland Maths [10M], teachers tackle the problem of estimating how many people can stand on a football (soccer) pitch. They don't do a very good job, largely as a result of not knowing how big a soccer pitch is, or how large their chosen unit (the foot) is. They start by assuming that the pitch is about 22 feet long (this is less than 7 metres), then that it is about 1/3 as wide as it is long – at which point one comments that the goal is at least 8 feet wide so that the width is wrong. Then they backtrack and start reasoning from the size of the goal relative to the height of the player to get a better estimate. It is the need to develop this ability to spot an implausible consequence of a mistake or poor assumption – not the calculations involved – that makes these 'plausible estimation' activities valuable in a world where technology has immense computational power but zero common sense.

The *Coin Counting* activity [10N] asks students to estimate the value of a pile of mixed coins being fed into an automatic coin counter. This is a perfect example of a scenario of where having a rough idea of the expected answer would help a user to spot any gross failure of the technology. Having a reasonable expectation of the cost of items in a trolley is one way to guard against scanning errors, such as scanning something expensive twice. Bowland Maths also includes a classroom project *You Reckon?* [10P] consisting of a collection of 'plausible estimation' tasks.

Beyond estimation, experienced users of any system perform a range of checks such as looking at max and min values in a set or keeping tabs on the number of entries with a given property.

10.7 Units and orders of magnitude in technology

Computer technology in the twenty-first century has introduced to everyday life a slew of new measurements and units. These were once only relevant to engineers and technicians – now they figure into common consumer decisions. Pick any advert for an IT product or service and there is probably some use for mathematical literacy.

ACTIVITY

The following was part of an advertisement for broadband internet:

Fast average download speeds up to 66 Mb.
Great for streaming, video calling, and gaming on multiple devices.

- Does it make sense?
  (No – 66 Mb is a quantity, not a speed – they mean 66 Mb/s.)
- Does an average speed of 66 Mb/s guarantee "great streaming"?
  (No – streaming video or music needs a guaranteed *minimum* speed.)
- How long would that take to download a 1 TB file at average speed?
  (Be careful – the lowercase 'b' usually means 'bit', the uppercase 'B' usually means byte normally 8 bits, so there's at least a factor of 8 there.)

The growth of computing power – including memory and storage size – has increased exponentially over the last 50 years or so. In fact, Gordon Moore (co-founder of Intel) speculated in 1965 that the number of transistors that could be fitted on an integrated circuit (then a major limiting factor in the memory capacities of computers) would double every two years – literally (really) an exponential growth – a prediction that has proven to be surprisingly accurate, possibly because the industry saw it as a target rather than a prediction. So, now, when buying a new laptop, you need to know the SI prefix for 10^{12} (T or tera). The amount of data generated worldwide in 2022 was predicted to be 97 zettabytes (Statista, 2021): 1 zettabyte = 10^{21} bytes. To deal with this growth in large numbers, two new SI prefixes have been adopted in 2022: ronna (R) is 10^{27} and quetta (Q) is 10^{30} – although quettabyte memory sticks are unlikely to be showing up in the shops just yet. The much wider range of numbers and units now required were identified in Section 3.3 as one way in which mathematics curricula must adapt for mathematical literacy.

Image sizes and resolutions

Computer imaging – now part of every mobile phone – brings more issues of scaling and dimensionality into the scope of consumer decisions. Converting between the measures used in advertisements and the details you need to make decisions can be tricky, for example:

ACTIVITY

Your old phone has an "8 Megapixel" camera. You look at your photos on a "Full HD" TV which has a vertical resolution of 1,080 pixels and an aspect ratio of 16:9. If you want to see more of the detail in your pictures, would you

a buy a new "4k UHD" TV which has a *horizontal* resolution of 3,840 pixels. (Yes, beware, between "HD" and "4k", they changed the definition of TV resolution from vertical pixels to horizontal pixels!) *or*

b buy a new phone which has a "12 Megapixel" camera.

Whatever – you want a new TV anyway! You're looking at one with a 120 cm diagonal screen and a "16:9 aspect ratio". The rim around the screen is pretty small, less than 1 cm.

Will it fit in an alcove 110 cm wide? Maybe create a spreadsheet (using formulae) so that you can work this out for other TVs you're considering.

A mathematical literacy investigation of image resolution could be extended to include factors like how the amount of storage (or internet bandwidth) varies for different image resolutions.

10.8 Further opportunities

This chapter has only scratched the surface of the potential uses of computer technology in the classroom, picking out a few cases particularly relevant to mathematical literacy. Coding is another area that offers potential. As early as the 1980s, students were learning to code with LOGO (Papert, 1980) and it was immediately evident that this opened up many new ways of developing mathematical concepts and also to developing 'know about' appreciation of how computers work. There are now multiple child-friendly versions available, most notably Scratch [10Q] where students can program with movement, music, and colours. Many secondary school science departments will also have sets of digital sensors that can be used to record attributes such as position and speed, light intensity, pH, temperature, dissolved oxygen, and many others. The kits are also equipped with video and data analysis tools, which support students bringing their understanding of mathematics to enrich explorations of many real-world situations. Students could, for example, make their own videos and graphs to analyse situations like those in the basketball example in Section 3. Digital sensors used to gather and analyse data are ideal to support interdisciplinary work across the health and science subjects.

Unsurprisingly, because the rise of information technology is a major factor in raising the demand for mathematical literacy, more examples can be found in the other chapters of this book. For example:

- Chapter 2: Mechanical linkages and deductive geometry lessons (Figure 2.4.2)
- Chapter 4: Using a simple app to illustrate the effect of random variations in the frequency of low-probability events (Figure 4.2.4)
- Chapter 4: Epidemic spreadsheet (Figure 4.4.1)
- Chapter 6: Creating Gannt charts, using graphing software for linear programming (Figures 6.3.1 and 6.3.2)
- Chapter 8: Finding and presenting demographic data
- Chapter 9: Financial planning using spreadsheets (Figure 9.1.1), Modelling interest with spreadsheets (Figure 9.3.1), 'The investment game' (Figure 9.4.2)

Acknowledgements

Figure 10.2.2 from Bowland Maths appears courtesy of the Bowland Maths maintainers. Figure 10.3.1 appears in Ridgway et al. (2007) which is available under a Creative Commons Attribution licence – the original task was designed by Jim Ridgway, Daniel Pead, and others as part of the *World Class Tests* project.

Figures 10.3.2 and 10.3.3 appear courtesy of *reSolve: Maths by Inquiry* [10D] – *reSolve* materials are available under the Creative Commons Attribution-NonCommercial-Sharealike 4.0 International (CC BY-NC-SA 4.0) licence.

Figure 10.4.1 appears courtesy of Dan Meyer at desmos.com [10E].

Figure 10.5.1 was captured from the World Bank website using data from UNESCO as cited in image. Figure 10.5.2 includes World Bank and UNESCO data licensed under the Creative Commons Attribution License 4.0.

References

Asp, G., Dowsey, J., Stacey, K., & Tynan, D. (2004). *Graphic algebra: Explorations with a graphic calculator*. Key Curriculum Press.

Babbage, C. (1864). *Passages from the life of a philosopher*. Longman and Co.

Hoyles, C., Noss, R., Kent, P., Bakker, A., & Bhinder, C. (2007). *Teaching and Learning Research Briefing Number 27*. https://discovery.ucl.ac.uk/id/eprint/1515625/

Papert, S. (1980). *Mindstorms: Children, machines and powerful ideas*. Basic Books.

Pierce, R., & Stacey, K. (2006). Enhancing the image of mathematics by association with simple pleasures from real world contexts. *Zentralblatt fur Didaktik der Mathematik*, *38*(3), 214–225.

Pierce, R., & Stacey, K. (2011). Using dynamic geometry to bring the real world into the classroom. In L. Bu & R. Schoen (Eds.), *Model-centered learning: Pathways to mathematical understanding using GeoGebra* (pp. 41–55). Sense Publishers. https://link.springer.com/book/10.1007/978-94-6091-618-2

Ridgway, J., Nicholson, J., & McCusker, S. (2007). Reasoning with multivariate evidence. *International Electronic Journal of Mathematics Education*, 2(3), 245–269. https://doi.org/10.29333/iejme/212

Statista (2021). *Volume of data/information created, captured, copied, and consumed worldwide from 2010 to 2020, with forecasts from 2021 to 2025*. Retrieved 12 July 2023 from https://www.statista.com/statistics/871513/worldwide-data-created/

Weinzierl, T. (2021). The pillars of science. In *Principles of parallel scientific computing: Undergraduate topics in computer science*. Springer. https://doi.org/10.1007/978-3-030-76194-3_1

Links to useful material

To visit any of these links, scan this QR-code or visit *ltml.mathlit.org* – append the link code to go directly to the entry – for example, ltml.mathlit.org/10A

The original source links are given below for attribution purposes:

[10A] *Making and Selling a Magazine* – Bowland Maths
https://www.bowlandmaths.org.uk/materials/pd/online/pd_04/pdf/pd_04_handout_4.pdf

[10B] Video of Spreadsheet Lesson – Bowland Maths
https://www.bowlandmaths.org.uk/materials/pd/online/pd_04/pd_04_follow.html – Activity 3

[10C] *Oxygen* activity designed by the Shell Centre for the World Class Arena project
https://ltml.mathlit.org/10C

[10D] *Bicycle Cornering* – reSolve: Maths by Inquiry
https://www.resolve.edu.au/mathematical-modelling-cornering?lesson=1671

[10E] *Will It Hit the Hoop?* – desmos.com
https://teacher.desmos.com/activitybuilder/custom/56e0b6af0133822106a0bed1?collections=5e827a6e58f1e36e4d220ef8

[10F] *Habitat Modelling Following Fire* – Victorian Department of Education
https://www.education.vic.gov.au/school/teachers/teachingresources/discipline/maths/Pages/numeracy-for-all-learners.aspx

[10G] *100 Years of Habitat Change: An Animated Fire Ecology* – Ian Lunt Ecology
Direct link: https://ianluntecology.com/2014/05/25/animated-fire-ecology/
Video link: https://www.youtube.com/watch?v=30j1tPc-m1s

[10H] Sources of demographic data – various
https://data.worldbank.org/
http://uis.unesco.org/
https://www.gapminder.org/resources/
https://www.cia.gov/the-world-factbook/

[10J] *Educational Datasets* – CSIRO (Australia's national science agency)
https://www.csiro.au/en/education/resources/educational-datasets

[10K] Graph of Literacy Rates
https:/ltml.mathlit.org/10K

[10L] *Fat Finger Error* – Wikipedia
https://en.wikipedia.org/wiki/Fat-finger_error

[10M] *Teachers Trying a Plausible Estimation Problem* – Bowland Maths (2008)
https://www.bowlandmaths.org/materials/pd/online/pd_02/pd_02_intro_01_footy.html

[10N] *Coin Counting, by Dan Meyer* – 101 questions
https://www.101qs.com/3199

[10P] *You Reckon* – Bowland Maths (2008)
https://www.bowlandmaths.org/projects/you_reckon.html

[10Q] Scratch Coding Language & Community – Scratch Foundation
https://scratch.mit.edu/

11

THE IMPORTANCE OF CURIOSITY

What are the attitudes and skills that lead a person to ask themselves the kinds of questions that we have explored in this book? What makes a productive disposition towards being mathematical literate? At the heart of this attribute is *curiosity*. The generic question is:

What is going on here?

Unless and until a person asks this question, the thinking that is mathematical literacy, beyond the purely functional, will not begin. Of course, the same question is at the heart of doing, rather than just learning, mathematics itself.

In Chapters 4–9 we raised such questions across six domains of real-world importance. We now want to explore in a little more depth the nature of curiosity and how it may be developed – by individuals and in the classroom. In Section 11.1 we sketch the history of thinking about curiosity in philosophy and psychology, moving on to education and mathematical literacy in Section 11.2. Section 11.3 briefly reviews the range of things that stimulate curiosity, moving on in Section 11.4 to examples of lesson design strategies for developing curiosity in the classroom. Section 11.5 looks at what we can learn from a quite different approach that comes from art and design.

In terms of the context-focused mathematics framework of Chapter 1, curiosity is a central, but often overlooked, part of the component 'demonstrating a productive disposition'. A person has a productive disposition for mathematical literacy when they expect that using mathematics will be helpful and informative, and have confidence that they will be able to use their knowledge to get a useful result. Productive disposition also requires a willingness to persist through obstacles, to be prepared to think carefully and

DOI: 10.4324/9781003303503-12

work flexibly exercising initiative. Curiosity about phenomena is an essential precursor of this kind of thinking.

11.1 What is curiosity?

Everyone is curious from time to time, wondering about something that happens, even if not actually formulating questions about it and seeking answers. Some people seem naturally more curious than others. In this section we explore the nature of curiosity, and how it fits into the broader domain of 'information seeking'.

Curiosity has long attracted the active interest (curiosity?) of philosophers, psychologists, neuroscientists, and other thinkers. William James (1899, republished 1983) called curiosity "the impulse towards better cognition". Daniel Berlyne (1954), an important figure in the twentieth-century study of curiosity, distinguished *perceptual* curiosity from *epistemic* curiosity. Perceptual curiosity is seeking out novel stimuli, the primary driver of exploratory behaviour. It is found in non-human animals and human infants, and is one driving force of human adults' exploration. Epistemic curiosity seeks to obtain access to information capable of dispelling uncertainties of the moment, but also to acquire knowledge. Other perspectives abound. Yet the Kidd and Hayden (2015, p. 449) in their review of the history concludes: "Despite its pervasiveness, we lack even the most basic integrative theory of the basis, mechanisms, and purpose of curiosity". However, the most popular theory about the *function* of curiosity is to motivate learning – hence this chapter. We adopt a heuristic approach, leaning on results of research where they seem directly relevant to curiosity's role in learning and teaching for mathematical literacy.

Some people have found it useful to distinguish different kinds of curiosity, related to the information being sought. *Information gap filling* or *fact seeking* is often routine, involving little curiosity, though they are important in *resolving ambiguity, dissonance*, or even *suspense*. The curiosity that underpins mathematical literacy is usually focused on looking for *patterns* and seeking *explanations*.

It is encouraging that research on young children confirms every parent's experience that curiosity is innate in all of us. Infants explore the world around them from the very beginning, making their own decisions, insofar as they can, on what to explore. They are curious learners whose cognitive development involves imposing structure on the environment they explore. Piaget, a pioneer of such studies, called young children "little scientists". More than half a century of research in both lab and more realistic settings has developed a rich picture of the curiosity-driven investigative processes involved in early-years cognitive development. This research has brought about a broad consensus that infants' information selection and subsequent learning in empirical tasks are influenced by their current thinking, the stimuli in the learning environment, and discrepancies between the two – that is, the element of surprise (for a review, see Mather, 2013). They are asking: "What is going on here?"

As children get older, their curiosity naturally moves with the things that most interest them in their everyday life. For some it may be a particular sport – news about the teams, the players, the results – or another hobby, including online games, or just things that occur in their family or their social life, extended greatly through the world of social media. This continues for all of us into our adult lives. But some people show more curiosity than others.

Some educational designers seek to transfer this curiosity, interest, and enthusiasm to motivate learning in school. This Philosophers' Stone of motivation has proven elusive, at least in terms of penetrating mainstream curricula – but work continues in the ways described below.

11.2 Curiosity and education

It has often been asserted by various reformers that formal schooling drives their innate curiosity out of many young children. It is true that traditional curricula, particularly in mathematics and science, focus on having students learn established knowledge – selected, demonstrated, explained, and illustrated by the teacher. Teacher and student questioning are focused on the understanding of these explanations, and how to perform skills to answer imitative practice exercises.

It is equally true that this approach has long been challenged, often under the banner of social constructivism. Pioneers such as Dewey (1916) described learning as fundamentally an exploratory process built on experience in learning. Gattegno (1970) saw the priority to be learning rather than teaching (reflected in the TRU framework set out in Chapter 3). He emphasised that learning takes place in stages – from 'awareness' that there is something to explore through a process of deepening exploration until that aspect no longer needs attention. In the UK this inspired the formation in the 1950s of the Association of Teachers of Mathematics, which promotes an approach to learning mathematics that is based on investigation of a wide variety of mathematical 'microworlds', guided by the teacher but with as much student autonomy as the particular young person can handle successfully. Leading associations of mathematics teachers around the world promote a similar philosophy and support teachers to implement it. As we saw in Chapter 10, information technology has a huge potential for stimulating and sustaining curiosity through investigation, one emphasised by Seymour Papert (1980) who created the Logo programming language for that purpose, with Turtle Geometry as a core example.

Curiosity is central to investigation: the process of asking what is going on in each problem situation. It involves the student turning a broad question into a sequence of ever more specific questions related to the problem situation, while monitoring their own progress (or lack of it) to see if a change of approach might be more productive (see Schoenfeld, 1985). Most coherent attempts at improving mathematics education have seen posing such questions, as a key component of doing and learning mathematics. *The Art of Problem Posing* (Brown & Walter, 2004) explores this issue within mathematics in readable depth.

This has been carried forward in the recognition, since the 1980s, of the importance of students tackling non-routine problems, in which curiosity is central, as part of their mathematics education. Building on the introspections of Polya (1945), an empirical tradition of research and development in 'problem-solving' has led to further insights and to well-engineered teaching materials – see, for example, *Mathematical Problem Solving* (Schoenfeld, 1985), *Strategies for Problem Solving* (Stacey & Groves, 1985, p. 2010), *Thinking Mathematically* (Mason et al., 2010), *Problems with Patterns and Numbers* [11A], *Classroom Challenges* [11B], *Bowland Maths* [11C] and *reSolve* [11D]. Such materials enable all teachers who are working in supportive environments to build mathematical problem-solving into their enacted curriculum.

Curiosity in mathematical literacy

Investigation, and thus curiosity, is at the core of mathematical literacy. Curiosity about real-world phenomena is an essential precursor to seeing how mathematics can help one understand them. How does an investigative approach to mathematics teaching generalise to situations in the real world? The problem situations, tasks, and activities in previous chapters are designed to demonstrate this – developing mathematical literacy through 'question posing' that builds towards 'insight seeking'.

In the real world (as in genuine investigation into pure mathematics), problems rarely arise as neatly posed questions for you to respond to; you yourself have to recognise a problem situation and generate specific questions as a starting point for further investigation. The stimulus is often a more or less open problem situation provoking you to ask: "What is going on here?"

For example, *Table Tennis Tournament* (first presented in Figure 2.1.2 and discussed more fully in Section 6.1) asked the open questions:

> Three of us have agreed to organise a table tennis tournament for our club. How should we set about it? What information will we need?

A group of students who are used to exploring open problems will identify relevant issues and associated variables, listing more specific questions, as discussed in Section 6.1, some of which are basic (how many tables are available) and others involve significant mathematical thinking. For example, considering whether to have a knock-out tournament, or a 'round robin' league where every player plays every other, leads to the question: how many games does each format require for a given number of players?

The answers are $(n-1)$ games in a knock-out versus $n(n-1)/2$ for n players in a round robin league. This shows that a league format requires either only a small number of players or a long time, which explains the choice of knock-out tournaments for professional tennis, and for ancillary tournaments in professional football where the typical league takes most of a year. In Chapter 6 we looked in more detail at the demands and challenges this task presents to students and teachers.

11.3 What stimulates curiosity?

In this book we have chosen situations that can excite curiosity in classrooms. Risk (Chapter 4) is an area that features in the press and other media with stories that may or may not warrant the concern implied. On planning (Chapter 6), we all know that we should probably be more systematic about it. We recognize that what we are told in politics and commerce is often misleading – and no one likes 'to be made a fool of' (Chapter 7). Much the same applies to the ultimately quantitative domain of money (Chapter 9). We hope the specific cases we have looked at in each of these areas will excite curiosity. But, as with so much in education, the real measure of success is how students and the adults they become can apply the elements they have learned to new situations. So what are the elements of curiosity and how can they be developed in classrooms?

We have noted that curiosity is a natural attribute of human beings from their earliest years. Nonetheless, people vary greatly in how far they want to ask questions and explore issues or simply accept the world as it is, something we all have to do to a great extent – you can't investigate everything. Here we ask what are the attributes that tend to make people curious. In Section 11.4 we discuss how these attributes can be developed, in the classroom and outside.

Surprise

Unexpected events are a common cause of curiosity. "What was that?" and "Look at that!" are natural immediate responses to surprising happenings and objects. Social media are full of videos of things– some genuine, some faked – that stimulate curiosity, and attract 'likes'. But surprise is not always turned into active investigation. And indeed, becoming surprised relies on some knowledge of what is normal – not surprising. Nonetheless, deliberately trying to generate surprises in the classroom is a powerful lesson design tactic. If surprises arise in the course of thinking things through, so much the better.

ACTIVITY

Small groups exchange examples of things related to mathematical literacy that have surprised them – and whether they were, or seemed, worth investigating.

Concern

Though extreme anxiety can cause paralysis, when a concerning or threatening situation arises, most people are moved to look into it – to begin to ask questions. If your landlord moves to raise your rent, you seek information on what safeguards there may be in your city. News reports about climate change raise obvious questions: how much global warming should we expect, when and with what effects, and what can we do about it? The 'risky' situations we discussed in Chapter 4 are examples where concern inspires curiosity.

The next step may be data-driven and/or theory driven. First, there arise issues of data reliability. Should I believe the climate scientists or the oil companies who assure the public 'it is all under control'? The second kind of follow-up is to try to *think through* the worrying phenomenon in a scientific way. On global warming, how can an increase in average temperature of just 2 degrees Celsius, which one would hardly notice on a summer day, have such profound effects? This may lead to a search for scientific explanations of the mechanism(s), or broader evidence that past predictions are proving correct, of the kind in Chapter 5.

ACTIVITY

Small groups look for real or potential sources of anxiety and discuss what steps they might, or might not, take as follow-up – noting the questions they would ask.

Opportunity

A more positive stimulus for curiosity arises when new opportunities are perceived. "Wanted, enthusiastic young people for engineering apprenticeships". "Expedition to the Amazon seeking skilled canoeist". This kind of message aims to stimulate the curiosity of good candidates for the task.

But the most obvious area of opportunity is money. We are bombarded with 'irresistible offers' – in print, on television, on social media, and even on our phones – hawking diverse ways to make money. Those who launch such enticements aim to stimulate curiosity but, often, to *avoid* closer investigation. Many sellers are highly skilled at the latter, even – or perhaps particularly – for purchases of the worst kind, from substandard products to scams, harmful or illicit goods. Mathematical literacy, with its emphasis on both critical thinking and mathematical tools, aims to encourage and help people look more closely. Basic advice like "If it seems too good to be true, it probably is" is a start but deeper investigation as we discussed in Chapter 9 is needed to evaluate the risk–benefit balance that underlies good decision-making.

ACTIVITY

Ask small groups to discuss how they react to approaches that purport to offer money-making opportunities, including the aspects that help them decide to move forward, or reject.

Success

An activity that has gone well leads one to persist, to look further into it, or to seek similar experiences. Among well-known indicators of positive attitudes and productive disposition is "persists in the activity after the allotted time". Every teacher or professional development leader remembers the times when "they kept going after the bell went" or "the questions and comments kept coming so we nearly missed lunch".

ACTIVITY

Ask for students who are striving for excellence in an activity to say something about the ways in which they motivate themselves to persist.

ACTIVITY

Stephen Anderson said "Surprise can be a very minor change that adds flavour and variety to an otherwise routine experience". Is that true? For you?

Knowledge

Beyond any immediate surprise, you are unlikely to be curious about something that you know nothing about. Loewenstein (1994) suggested that a small amount of information serves as a priming dose, which greatly increases curiosity. Consumption of information is rewarding but, eventually, when enough information is consumed, satiation occurs and later information serves to reduce further curiosity.

More broadly, a good 'seed' for engendering curiosity, and subsequently mathematical literacy, might be one of an individual's current interests, about which they already have some 'primer' knowledge and a motivation to understand more, and which has some authentic mathematical content that can be 'surfaced'. For teenagers such subjects are likely to include social media algorithms, social media dynamics ("likes" generation, dynamics of viral posts), fashion cycles, dynamics of communication and relationships, climate change, broad social concerns to build a better world, e-sports, and even the mathematics of exams.

Curiosity within information seeking

How does curiosity relate to information seeking in the broad sense? Are they the same? This question has been central to all theoretical attempts describe and classify curiosity – with the absence of any agreed model. We take the view that it is most useful to see curiosity as a form of information seeking *where an element of novelty is involved.* However, persisting in an information-seeking activity, even after the minimum information required to satisfy the immediate need has been found, indicates deeper curiosity. For example, in areas we know a lot about, professionally or informally, we are likely to want 'to keep up to date'. Some of this is *routine* information seeking but going further involves curiosity.

What the different kinds of stimuli described above have in common is that they are based in *emotion.* Curiosity is an initial response that may lead to an intellectual one – fast thinking perhaps leading to the slow thinking that mathematical literacy, mathematics, and all intellectual activity involves. In Section 11.4 we discuss the pedagogical challenges involved – both in stimulating curiosity in students and in encouraging the transition to analytical thinking.

11.4 Developing curiosity in the classroom

We noted in Section 11.1 that the most popular theory about the function of curiosity is to motivate learning. From a teaching perspective, there are two rather different challenges:

- How can we support students who are interested in understanding what is going on?
- How can we stimulate the curiosity of students who see learning as no more than understanding (or even simply remembering) the things they have been told?

Though there is clearly an overlap, we will distinguish these as *feeding curiosity* versus *developing curiosity*. Over recent decades, a lot of progress has been made in answering the first of these questions. The second challenge remains work in progress. Some lesson designers tend to assume that all students are naturally curious and investigative which, to say the least, is empirically 'not proven'; teachers are less likely to make that mistake.

Feeding curiosity

Through a mixture of research and creative design over many years, lesson activities have been developed that support investigation of phenomena in the world beyond the classroom, as well as within mathematics. Teaching the modelling of everyday situations – an underlying theme of this book – has been going on at least since the 1960s when one of us (Burkhardt) introduced experimental workshops built around the following kind of questions:

- If you are crossing between buildings in heavy rain, do you get wetter walking or running?
- If you plan to buy a used car, what age of car should you buy and when should you sell it to minimise the cost?
- What ways are there in which climbing a ladder against a wall can become unstable and which are most dangerous? This example has been discussed in Section 6.4 to illustrate the climate discussion of tipping points.

Developing curiosity – drama in lesson design

In their article on design, Burkhardt and Pead (2020) suggested that "Every classroom is a theatre. Use that fact in lesson design – but make the teacher the director not the star" [11E]. This was inspired by the work of Dan Meyer (see below) and Malcolm Swan. Unlike some other subjects, mathematics lessons rarely have dramatic tension, a story, or a surprise. Usually, in theatre terms, the teacher is both director *and* star. The students are the audience. The play begins with a long soliloquy, then it's over! Students just imitate many times over what they have been shown. The drama design approach can increase student motivation and problem-solving in a broad sense.

As we have tried to show throughout the book, the task-based lessons inherent in mathematical literacy *can* have a dramatic or topical story inherent in the problem situation. (A fantasy context can also be contrived to introduce more abstract pure mathematical problems.) The key strategy is to engage the students by introducing an element of storytelling or role-play.

The widely admired designer Dan Meyer devised an explicit structure for whole-group task-based lessons with the title *Three Act Math* [11F]. The dramatic structure is based on a standard model for writing stories:

- *Act One* introduces the central conflict, an engaging and perplexing one (i.e. designed to excite curiosity).
- *Act Two* in which the protagonist (student) overcomes obstacles, looks for resources, and develops new tools.
- *Act Three* is a solution-discussion and solution-revealing phase that resolves the conflict and sets up a sequel (extension problem).

In the examples he developed, Act One is often a somewhat perplexing video setting the scene plus a question or two, made explicit by the teacher or left to emerge from discussion. For example, two 'back of the envelope' estimation tasks from Meyer's 101 questions website [11F] are Cola Pool and Coin Counting.

> *'Cola Pool'* shows the last bottle of the soft drink being poured into a cylindrical backyard pool and asks: *How many bottles of Cola did they buy to fill that pool?* The first question is *What information is needed to answer the question?* The information is then provided and the calculations proceed through to discussion.
>
> *'Coin Counting'* is more demanding – the video shows a pile of coins of various values. The questions are *How much cash is that? How many coins are there?* Here deciding the necessary information and how to find it is the core challenge.

The three-act lesson approach has spread widely in North America. Ontario, for example, which has a fine tradition in mathematics education, presents the three-act approach a little differently, with the goal to spark curiosity and fuel sense-making. They offer some interesting real-world tasks [11G] as does the Bowland Maths project in the UK [11C]. An example from Bowland Maths is "Reducing Road Accidents" (see Section 3.2) in which students are given a budget and make road safety recommendations to the town council. Experience shows that lessons with a dramatic structure like these seem to engage all students, not just the already-curious (TRU dimension 3).

The use of a story element that encourages curiosity can be found, to a greater or lesser extent, in the design of lessons on standard topics in mathematics. Leslie Dietiker (2015) and colleagues illustrated this with a detailed comparative analysis of the lessons from four different textbook series – here on the minimum information needed to establish that two triangles are congruent. The four lessons differed substantially in the time for which all the various combinations of angles and sides were left open as possibilities. For example, a lesson starting with open questions like "If all three angles are the same, must the triangles be congruent?" explores a wider space than only showing that the three standard sets of conditions (SSS, SAS, ASA) are sufficient. In summary, leaving more questions open for longer is a method of encouraging both curiosity and student agency and autonomy (TRU dimension 4).

11.5 Further thoughts

In this book, as in mathematics education more widely, we have adopted a largely scientific and utilitarian approach: how can mathematics help people navigate life and its challenges more effectively? A different approach to education that emphasises the value of curiosity has been suggested by Elliot Eisner (2002), a pioneer in arts education. In it more importance is placed on exploration than on discovery, more value is assigned to surprise than to control, more attention is devoted to what is distinctive than to what is standard. He further declares his preference for *aesthetic* ways of knowing and learning, pointing out that artists develop and achieve their ends through exploiting emerging and unexpected opportunities along the way. He rejects the common position, summarised as "science is useful; the arts ornamental". His vision of education as an art inspires new questions that serve to address some of its problems: "How can the pursuit of surprise be promoted in a classroom? What kind of classroom culture is needed?"

What might this mean for mathematics and for mathematical literacy? Mathematics has always had a strong aesthetic element. Concise 'elegant' proofs are more pleasing than 'proofs by exhaustion' that consider every case separately. Mathematical patterns and structures have immediate appeal for some people. Dietiker (2015) explores the new visions, values, and practices that taking aesthetics seriously might open up in mathematics education, informing a reimagining of mathematical experiences in classrooms. She sees the value of mathematics lessons designed as mathematical stories of exploration, extending this approach to the teaching of regular concepts. This echoes the concept of dramatic design in Section 11.4, and the mathematical literacy examples there and throughout the book, but with an explicit emphasis on the aesthetic pleasures and the curiosity that drives the search for insights involved.

In short, the processes of mathematical literacy should be enjoyable as well as informative.

Acknowledgement

We are grateful to Leslie Dietiker for informative conversations.

References

Berlyne, D. E. (1954). A theory of human curiosity. *British Journal of Psychology*, *45*(3), 180–191.

Brown, S. I., & Walter, M. I. (2004). *The art of problem posing* (3rd ed.). Taylor and Francis eBook. https://doi.org/10.4324/9781410611833

Burkhardt, H., & Pead, D. (2020). 30 Design strategies and tactics from 40 years of investigation. *Educational Designer*, *4*(13).

Dewey, J. (1916). *Democracy and education: An introduction to the philosophy of education*. The Free Press.

Dietiker, L. (2015). Mathematical story: A metaphor for mathematics curriculum. *Educational Studies in Mathematics*, *90*, 285–302.

Eisner, E. W. (2002). What can education learn from the arts about the practice of education. *Journal of Curriculum and Supervision*, *18*(1), 4–16.

Gattegno, C. (1970). *What we owe children: The subordination of teaching to learning*. Outerbridge and Diensfrey.

James, W. (1899). Talks to Teachers on Psychology, and to Students on some of Life's Ideals. *Science*, *9*(235), 90–910.

James, W. (1983). *Talks to teachers on psychology: And to students on some of life's ideals* (Vol. 12). Harvard University Press.

Kidd, C., & Hayden, B. Y. (2015). The psychology and neuroscience of curiosity. *Neuron*, *88*(3), 449–460.

Loewenstein, G. (1994). The psychology of curiosity: A review and reinterpretation. *Psychological Bulletin*, *116*(1), 75.

Mason, J., Burton, L., & Stacey, K. (2010). Thinking *m*athematically (2nd ed.). Pearson.

Mather, E. (2013). Novelty, attention, and challenges for developmental psychology. *Frontiers in Psychology*, *4*, 491.

Papert, S. (1980). *Mindstorms: Children, machines and powerful ideas*. Basic Books.

Polya, G. (1945). *How to solve it*. Princeton University Press.

Schoenfeld, A. H. (1985). *Mathematical problem solving*. Academic Press.

Stacey, K., & Groves, S. (1985). *Strategies for problem solving: Lesson plans for developing mathematical thinking*. Latitude Publications. Second edition 2006.

Swan, M., Pitts, J., Fraser, R., & Burkhardt, H., with the Shell Centre Team. (1984). *Problems with patterns and numbers*. Joint Matriculation Board and Shell Centre for Mathematical Education [11A].

Links to useful material

To visit any of these links, scan this QR code or visit ltml.mathlit.org – append the link code to go directly to the entry – for example, ltml.mathlit.org/11A

The original source links are given below for attribution purposes:

[11A] *Problems with Patterns and Numbers* – Swan et al. (1984) https://www.mathshell.com/materials.php?item=ppn&series=tss

[11B] *Classroom Challenges* – Mathematics Assessment Project
https://www.map.mathshell.org/
[11C] *Bowland Maths*
https://www.bowlandmaths.org/
Reducing Road Accidents: https://www.bowlandmaths.org/projects/reducing_road_accidents.html
[11D] *reSolve* – Australian Academy of Science
https://www.resolve.edu.au/
[11E] *30 Design Strategies and Tactics from 40 Years of Investigation* – Burkhard, Pead (2020)
https://www.educationaldesigner.org/ed/volume4/issue13/article53/#link12
[11F] *The Three Acts of a Mathematical Story* – Dan Meyer
https://blog.mrmeyer.com/2011/the-three-acts-of-a-mathematical-story/
Cola Pool: https:// www.101qs.com/3896
Coin Counting: https://www.101qs.com/3199
[11G] Problem-based Math Lessons & Units – *Make Math Moments*
https://learn.makemathmoments.com/tasks/

12

DESIGNING MEASURES

What is a measure? A measure assigns a number to a characteristic of an object or event. Familiar measures include length and weight, time, and money. A measure of 'intelligence' may simply be the assignment of a number to a set of responses to a test.

Measures are important. Peter Drucker's (1995) assertion about macroeconomics is as follows:

> If you can't measure it, you can't manage it.

Drucker's assertion surely has limitations. How many of the decisions you make in your personal life depend mainly on quantitative measures? But we have seen how important measures, with the choice of variables that they include, are in understanding real-world phenomena and in making large-scale decisions. We have seen a lot of measures in this book: the probabilities of various events in Chapter 4 measuring risk, the atmospheric concentration of CO_2 and its relation to global temperature (Chapter 5), the length of the critical path in planning processes like *Airplane Turn-round* in Chapter 6, the 'goodness' of washing machines in Chapter 7, and many more.

Other familiar measures include the energy ratings for electrical appliances, various labelling systems for the healthiness of food products, and the Body Mass Index, used as a measure of obesity, defined as

$$\text{BMI} = \frac{\text{mass in kg}}{(\text{height in metres})^2}$$

Note that the dimensions of BMI are mass/length^2 or density × length so, assuming the density and shape of human bodies is roughly constant, the key

DOI: 10.4324/9781003303503-13

variable in BMI has the dimensions of a length. Waist measurement, another widely recommended measure, reflects this directly. The strengths and limitations of these measures are widely discussed.

How do we ensure that a measure captures the essential features we want? For some things it seems straightforward. Money is an obvious example: the amount of money in your bank account is important and well-defined. Though, even with money, our primary concern may be the 'buying power' of that money which changes with time. As discussed in Chapter 9, the well-defined rate of interest on a savings account is less important from a longer term perspective than the *real* rate of return after inflation is taken into account. That requires devising a measure for inflation as discussed in Section 12.3. For other abstract and social constructs, the amount of inequality in a country, for example, as in Chapter 8, it is far from obvious how a measure should be defined.

Creating definitions is an important aspect of doing pure mathematics, but often they are just stated and then learned definitions. Interesting fundamental *choices* in designing a definition, that give insight into the reasons for rules, can be all too easily passed over. Why, for example, is the product of two negative numbers defined as positive? The compelling reason is to make the distributive law work for negative numbers, so that the usual arithmetic of positive numbers continues to work in the extended system. For example, both of these calculations give the answer 1 when 'a minus times minus equals a plus'.

$$(4-3)(3-2)=1\times1=1$$

and

$$(4-3)(3-2)=4(3-2)+-3(3-2)=12-8-9+6=1$$

In this chapter, however, we look at the *design* of measures of abstract or social quantities of real-world interest that go beyond physical measurement. Constructing such measures is in fact a challenge of mathematical modelling: identifying the critical variables and choosing a function that combines them to create a useful number. First in Section 12.1, we focus on some classroom activities involving the design of measures, going on in Section 12.2 to a more general discussion of desirable properties of a measure. Finally, in Section 12.3 we look in more detail at the mathematics of some of the important measures we have discussed in earlier chapters.

12.1 Designing measures in the classroom

We begin with some tasks inspired by the Balanced Assessment Project design group at the Harvard Graduate School of Education, which called them '*-ness tasks*'. See Guadagnoli (2000) for examples of the Harvard *Balanced Assessment* team's work.

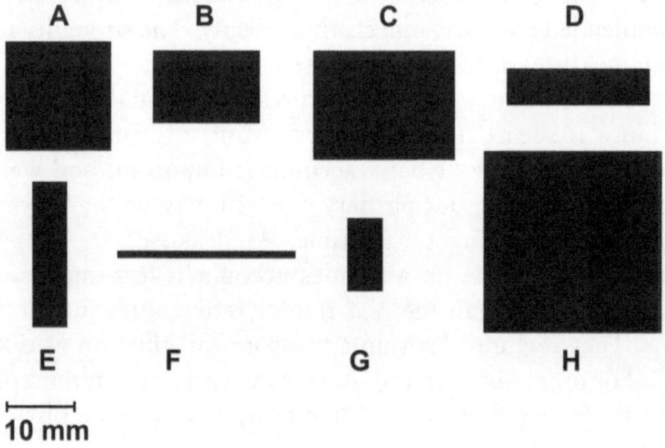

1. Order these rectangles from 'least square' to 'most square'.
2. Now design a formula for a measure of 'squareness'.
3. Compare your measure with other students' formulas. Which do you think is the better measure and explain why.
4. Try to devise a better measure.

FIGURE 12.1.1 'Squareness' task.

Source: Inspired by a Harvard GSE Balanced Assessment task [12B].

In Figure 12.1.1, students are asked to design a formula for a measure of 'squareness' S and compare it with their fellow students' suggestions. A possible discussion might be as follows:

> *Jo suggested the length (longer side) l minus the width w. $S = l - w$.*
>
> *Samira suggested the ratio $S = l/w$.*
>
> *Samira's measure only depends on shape; Jo's changes with size.*

Students are then asked to revisit and improve their work. For example:

> *Samira's measure, $S = l/w$. goes from 1 to infinity. Perhaps it is better seen as a measure of 'unsquareness'? An improvement might be $S = w/l$ which goes from 0 (super-thin, a line) to 1 (exactly square). More complex possibilities could be devised, especially when extending to shapes that are not all rectangles.*

The original *Balanced Assessment* materials include several tasks in this genre: *disc-ness* is similar to *square-ness* but looks at three-dimensional cylinders and *compact-ness* compares clusters of dots These tasks are abstract and 'pure' but help introduce some of the concepts about what makes a good measure.

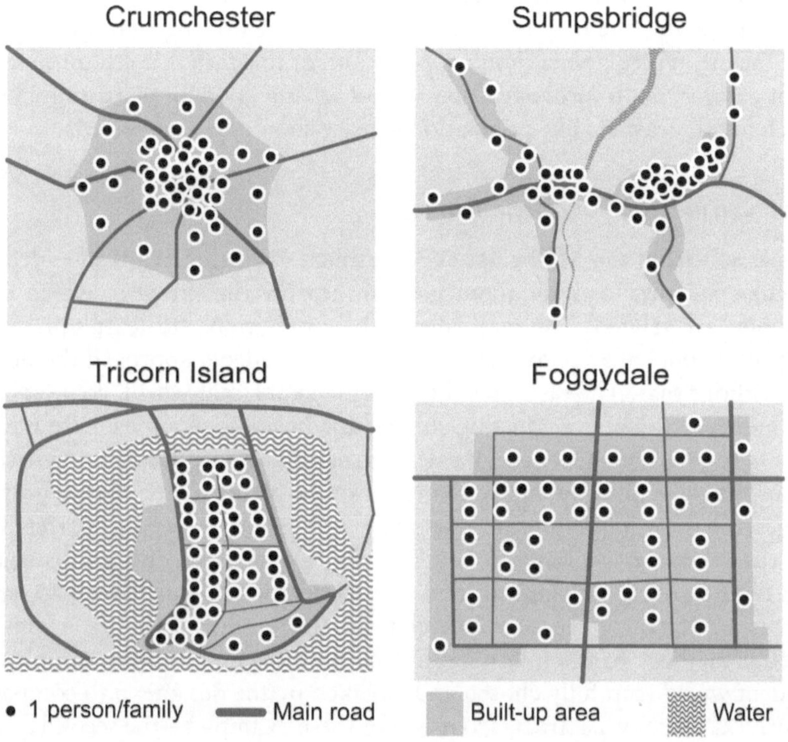

The four maps of imaginary towns above are all drawn to the same scale (about 5 km wide) and have about the same population.

Each dot on the map shows where one person or family lives.

1 Look at the maps and put them in order of how "crowded" you think they look.
2 Now design a method for calculating a measure of "crowded-ness".
3 Compare your measure with other students' work. Which do you think is the better measure and, most important, explain why.
4 Try to devise a better measure.
5 Extended task – find the population of some real medium-size towns and cities. How many people might one of the dots on the map above *really* represent? Look at satellite photos online and try to guess where most people live – could you estimate a value for your crowded-ness measure?

FIGURE 12.1.2 A more practical *-ness* task [12B].

The task in Figure 12.1.2 (and [12B]) is inspired by Harvard's *Compact-ness* idea but with a slightly more realistic context. A possible solution might be to calculate the average population density by dividing the number of dots by the estimated area of the built-up zone ... or should that be the area of the map (which would be the same for all four towns) or the 'bounding box' of the dots? Or is a town with large clumps of dots 'more crowded'? Perhaps you could divide the map up into a grid, count the dots in each square, and take the mean/median/maximum…? The aim of this task is not to produce a 'right answer' but to provoke discussion of what a good measure is and how widely used measures like population density should be interpreted.

Creating a measure for correlation

This lesson from the Mathematics Assessment Project [12C] is intended to provoke students to think about how to measure the extent to which two variables are related. The mathematics of correlation and least squares is the professional answer to this question: these students approach the problem without that. The task uses the semi-plausible 'real-world' context of a movie theatre owner conducting surveys for business planning – ice cream sales versus temperature, snack sales versus soda sales, number of viewers versus movie length. By happy chance, a quick glance at the scatter graphs suggests that the 'data' for one is strongly correlated, one weakly correlated, and the other is fairly random. Students are first asked to critique a suggestion of drawing a polygon around the scatter points, calculating its area and using 1/area to measure 'correlation'. They are then asked to devise a better method. In the follow-up lesson, they are given three examples of 'student work' (carefully chosen and tweaked by the designers) to compare. Figure 12.1.3 shows extracts from two of these examples – the second could be used to lead into teaching 'least squares'.

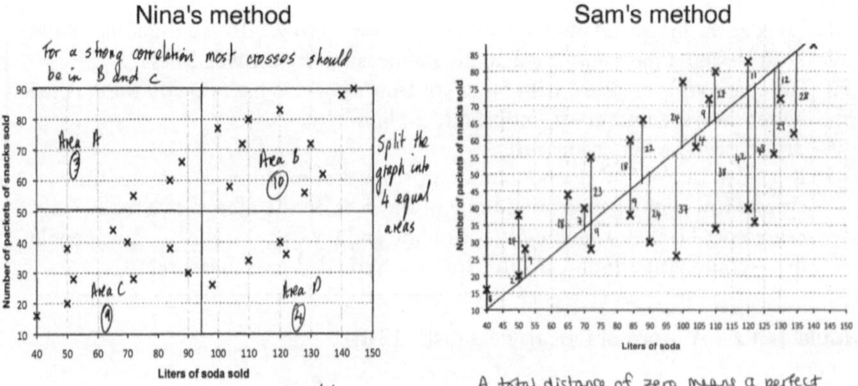

FIGURE 12.1.3 Sample 'student' work used in the MAP *Design a measure* lesson. Mathematics Assessment Project [12C].

12.2 What makes a good measure?

Even within the examples above, where the situation is simple and clearly defined, there are design issues worth discussing. For more complex life-related situations, as we have seen in previous chapters, these issues can be profound. Here we stand back from particular contexts and take a more general look at the properties that characterise a good measure, illustrating these with examples from previous chapters.

A measure is stronger if it

- quantifies an important feature of the situation of interest,
- is well-defined and explained so that anyone can use it,
- takes values that reflect the extent of the feature in a way that seems natural, and
- uses data that is easy to collect.

BMI, for example, fulfils these criteria; that does not eliminate its limitations.

There are also some other, less obvious, properties that are important. For example, no widely adopted measure has just the use, and the influence, for which it was designed – at the least, there are usually unintended consequences. Indeed, one rather general problem is summarised in Goodhart's (1975) Law, usually simplified as follows:

> When a measure becomes a target, it ceases to be a good measure.

This may be because people 'game it' – manipulate their activity pattern to help meet the target, which may be good but may involve suppressing other important activities not included in the measure. This is a design issue – some important aspects of the real phenomenon being measured have been overlooked or undervalued by the measure. Since the whole point of a measure is to reduce a complex phenomenon to an easily compared number, this is almost inevitable to some extent. Goodhart was originally critiquing monetary policy, but the principle extends to many other fields. If you need a minimum number of words in your essay, waffle. If you're a call centre operator 'scored' on how many calls you answer per day, get callers off the line as quickly as possible without solving their problem. If you want to increase your rating on social media, post lots of messages that a lot of people will 'like' and so 'upvote'. Criticising Android in an iPhone forum – or vice versa – is usually a banker. There are many examples of this in education, notably the effect of 'high-stakes' examinations on teaching, which we look at in Section 12.3.

To counter this, a good measure needs an explicit 'improvement mechanism' that evaluates the mechanism in use against the aims for which it was introduced. Price indexes, for example, are based on a 'basket of goods' that are representative of buying habits, so the goods included need to change with time. As we noted in Chapter 8, this makes comparisons over long periods increasingly approximate.

Log versus linear presentations

There are many issues about the presentation of data, which are affected by the design of the measure. Figure 4.4.4, for example, using data from early in the pandemic, presented the risk of dying from COVID-19 as a function of the patient's age (the key variable) in both linear and logarithmic form. The latter proved more informative because the risk changed by many orders of magnitude, so the logarithmic scale could show all parts of the graph clearly. As it happened in this case, there was an extra benefit from the logarithmically defined measure because, from middle-age upwards, the dependence on age was close to exponential, the risk doubling every eight years of age, so the graph is roughly a straight line.

Sometimes, logarithmic scales are a better model of human perception of phenomena. For example, the musical scale of octaves is a logarithmic one – each octave represents a doubling of frequency. Imagine listening to someone playing each note on a piano keyboard from the low bass to the high treble and sketching a graph of the 'pitch' you heard. You would probably perceive the pitch as rising steadily (so a straight line graph), rather than the frequency's exponential curve doubling every 12 notes. The decibel scale of loudness is also logarithmic, as is the Richter scale for the magnitude of earthquakes – both phenomena that vary over many orders of magnitude in the energy involved.

An unusual logarithmic measure features in Chapter 13 which considers how to compare the difficulty of guessing passwords creating using various rules – say, ten random alphabetic characters compared to a random English word or compared to a longer (but memorable) string of random English words. *Password entropy* uses a logarithmic scale to estimate the number of random binary bits (hard to crack, but impossible to remember) that would give the same number of permutations.

12.3 Some important measures – a critical review

This section is largely focused on mathematics. The societal issues are discussed more fully in earlier chapters. We start with an example from education, moving on to look at some important measures that we have discussed in earlier chapters, some including more advanced mathematics.

Unintended consequences of 'high-stakes' assessments

In many countries governments introduce compulsory assessments of student performances – examinations or tests – believing that they lead both to improved learning and to decision-makers making better informed decisions whether it be parents choosing a school or the government making policy decisions. Surely an admirable goal. "If you can't measure it, you can't manage it" again. However, they also want assessments that satisfy constraints:

- standardised – in the interest of fairness, surely essential,
- well-defined assessment objectives,
- don't take much "time away from teaching",
- don't cost much, and
- provide results that are clear and easy to understand (just one number preferred).

The normal outcome is assessments that, particularly in mathematics, are time-limited taking only an hour or two at most, consist of short 'items' that take only a few minutes to answer, with no extended chains of reasoning or student explanation. It is no surprise that teachers, whose career progress often depends on 'test scores', focus on task types with the limited range in the tests and often school leadership reinforces this. Burkhardt et al. (1990) summarised this as:

What you test is what you get (WYTIWYG).

Sadly, these assessments do not encourage many of the dimensions of learning that are recognised as important, particularly those involving substantial chains of reasoning or enquiry. While WYTIWYG has long been a familiar part of teachers lives, government and assessment providers have been reluctant to accept it as a fact. You hear phrases such as "We don't test this but, of course, all good teachers teach it" and "We're just sampling some important aspects of learning". This allows them to avoid the responsibility to provide 'tests worth teaching to' that sample all the ambitious goals for learning that are commonly listed in official curriculum documents. More recently, that has changed in some places with assessment objectives that, for example, separate skills, reasoning, and solving non-routine problems. But such progress in the rhetoric is, up to now, minimally reflected in the examination tasks themselves. Traditional tasks that demand extended chains of reasoning with explanations (commonplace as essays in humanities subjects) are now rare in mathematics assessment in some anglophone countries. Yet, as we have seen in this book, mathematical literacy largely involves such activities. PISA tasks are an invaluable but limited step in this direction.

Assessment does not have to be narrowed in this way. Past examples show (see, for example, Burkhardt, 2009) that even timed written examinations can cover a much broader variety of mathematical reasoning. Extended project work can be reliably assessed, although it takes time. Practical principles for the design of *High-stakes Examinations to Support Policy* (Black et al., 2012) were developed by a working group of the International Society for Design and Development in Education.

Picking your inflation measure: RPI versus CPI

There is a nice mathematical issue behind the measure of inflation currently preferred by the UK government – 'Consumer Prices Index' (CPI) – which, since 2003, has been gradually replacing the previous 'Retail Prices Index' (RPI) measure. CPI is the international measure, used by the OECD and governments to study inflation around the world. Both measures are correctly described as the percentage increase over the past year of "the average price of a basket of products and services" that is chosen to represent the spending of a typical citizen. (For the CPI, the actual goods and services are country-specific.)

One interesting mathematical difference between the two indices is that the 'retail price index' for inflation (RPI) uses the arithmetic mean of the increases across the basket while the 'consumer price index' uses the geometric mean of the increases. It is standard mathematics that the former is always larger – for two numbers, *a* and *b*, the difference of the squares of arithmetic and geometric means is the square of half the difference, which is always positive (or zero):

$$4\left(\left(\frac{a+b}{2}\right)^2 - ab\right) = (a+b)^2 - 4ab = a^2 + 2ab + b^2 - 4ab = (a-b)^2$$

So the difference between the results is influenced by the difference between the arithmetic and geometric mean of a sample, which in turn is proportional to the variance of the sample – illustrated by $(a-b)^2$ in the simple formula for two items above.

The UK government still uses both CPI and RPI. Interestingly, it uses

- the lower CPI for increases in *payments* such as pensions, unemployment and disability benefits;
- the higher RPI for increasing some *charges*, for example, train fares.

You may wonder why!

There are other important differences between RPI and CPI – notably the baskets do not contain exactly the same types of products and services.

RPI includes housing costs (rents, mortgage, payments) which CPI does not; CPIH, a version of CPI that allows for housing costs (by incorporating *yet another measure* – 'rental equivalence') is currently being introduced in the UK. A fuller discussion of RPI versus CPI can be found on the UK Office for Budget Responsibility's website [12D]. Despite their complexities, estimates of inflation are nonetheless essential to the modern world.

Choosing the 'basket of goods' becomes even more complex over time. Year-by-year an item 'drops out' as it becomes less significant in people's buying habits; another item is introduced. Such gradual change is non-controversial. But measuring inflation over decades, or even centuries as in Chapter 8, where the pattern of buying has changed qualitatively is much more difficult. The price of bread, for example, is much less significant in modern lives than it was long ago, and travel much more important.

How can you measure economic inequality?

This is not an easy question and has no unique answer. For instance, do you include wealth or only income? The measure used by Wilkinson and Pickett in *The Spirit Level* which we summarised in Chapter 8 is the ratio of the average of the highest 20% of incomes to the average of the lowest 20%; this varies in the data shown from below 4 in Japan to nearly 10 in the United States.

A more sophisticated measure of inequality is the *Gini Coefficient* [12E] – a number between 0 and 1 (though often expressed as a percentage) that is roughly the average unsigned difference in income between pairs of individuals in the population divided by the average income, that is, for n people:

$$G = \frac{\sum_{i=1}^{n} \sum_{j=1}^{n} \left| x_i - x_j \right|}{2n^2 \bar{x}}$$

The factor 2 arises because each pair is counted twice – each individual is both i and j in the n^2 terms. It follows that

- $G = 0$ if and only if everyone has the same income.
- $G \approx 1$ if one person in a large population has all the income.
- $G = 0.5$ if half of the population have the same high income, and the other half all have the same low income.

The Gini coefficient is not just for inequality of income. Clearly, wealth could be used instead of income, but is also useful in other domains. For example, Gini coefficients could be used to compare inequality in mathematics achievement between individuals within or between schools. Results from recent PISA studies (often using Gini coefficients) show that achievement

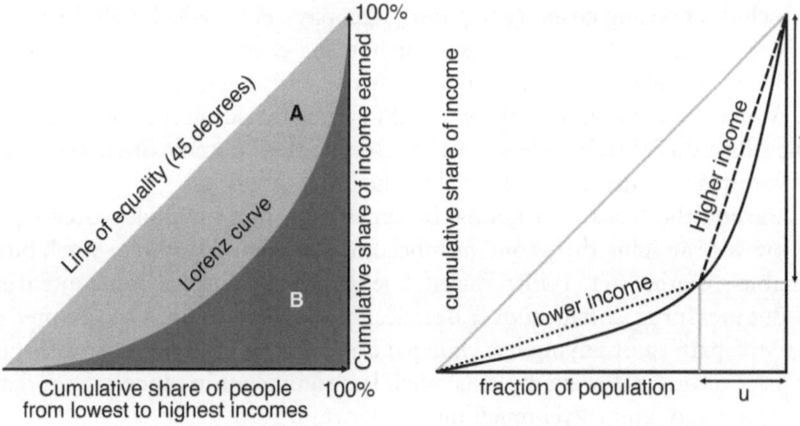

a. Gini coefficient G = Areas $\frac{A}{A+B}$

b. Population with just 2 incomes
G = f − u

FIGURE 12.3.1 Visual representation of the Gini coefficient and Lorenz curve in (a) general case and (b) special two income case.

Source: Adapted from illustrations on Wikipedia [12E].

is negatively influenced by academic and social segregation (inequality) between schools (OECD 2019), adding to the negative influence of individual inequality.

Another representation is through the Lorenz curve, which shows the cumulative income or wealth through the population, rank ordered from poorest to richest – illustrated in Figure 12.3.1. The slope of the curve represents the income at that point in the population. If everyone has the same income, the total grows linearly – the 45 degree line in Figure 12.3.1a. If one person has all the income, the curve runs along the horizontal axis and then shoots up the vertical axis. The curve shown represents a more typical population.

Figure 12.3.1b shows a situation where there are just two different incomes and the top u of the population has f of the income. The Lorenz curve consists of two lines – one for the $(1 - u)$ low paid folk, the other, steeper, line for the high paid. We can calculate the ratio of areas in this case giving

$$G = 2\left(\frac{1}{2}1 - \frac{1}{2}(1-u)(1-f) - u(1-f) - \frac{1}{2}uf\right) = f - u$$

A surprisingly simple result. (When one meets a complex calculation with a simple result like this, one should ask: "Can I find a simpler route to it?") Good to check that it fits the extremes cases: for equal incomes $f = u = 100\%$; $G = 0$; for one person with all the income $f = 1$, $u = 0$; $G = 1$.

12.4 The importance of units

Before we finish this chapter on measures, we should say something about the units we choose to express quantities in. Visualising and communicating the meaning of numbers effectively depends on understanding the units used. It can be made much clearer by using standardised units.

History tells a relevant story. Before the French Revolution of 1789, there were thousands of different units for length and weight in use across France. When most craftsman never went far from their village, this did not matter. The next century led to a government-led movement to standardise units in a 'metric system'. Initially physical objects – a standard metre bar and kilogram sphere kept in Paris – were used to define the unit. Subsidiary standards all over the country were then standardised against these – a cumbersome method. At a convention in Paris in 1875, other governments were persuaded to adopt the metric system.

Originally, physical 'standard' objects used had to be carefully maintained at national weights and measures bureaus. For example, until 2019, the kilogram was still defined as the weight of a particular lump of metal – the 'International Prototype of the Kilogram' held at the International Bureau of Weights and Measures in France. Countries had their own copies of this kilogram. The international standards movement has been working to redefine the standards (though not the size of the units) so that they can be reproduced anywhere. The second is based on the frequency of the light of a specific atomic spectral line and the metre is derived from that. Since 2019, the seven 'base units' of the SI system have all been redefined in terms of measurable physical phenomena. The best-known units are given in Table 12.4.1, more detailed information can be found from the National Physical Laboratory website [12F].

TABLE 12.4.1 Three of the base SI units

Unit	*Symbol*	*Description*	*Definition*
second	s	Unit of time	Defined as the time that elapses during 9 192 631 770 oscillations of light of a specific spectral line of the caesium 133 atom
metre	m	Unit of length	Defined by the value of the speed of light, 299 792 458 ms^{-1}, and the definition of the second
kilogram	kg	Unit of mass	Defined by the value of Planck's constant, $6.626\ 070\ 15 \times 10^{-34}$ kg m^2s^{-1}, and the definitions of 'metre' and 'second'

Source: Based on information from the National Physical Laboratory [12F].

The value of using standard units for making comparisons is now generally recognised. For example, the way that national income or GDP should be calculated is internationally agreed (though not universally observed by governments!) In other areas there are avoidable distractions – for example, during the COVID pandemic, death rates 'per hundred thousand people' were commonly used rather than per thousand or per million as international conventions encourage.

As we have noted (see, for example, Section 3.3), understanding very large and very small numbers is an important part of mathematical literacy – in particular the extending list of names for the various powers of a thousand: kilo, Mega, Giga, Tera, Peta going up and milli, micro, pico, femto going down. Especially driven by computing technology, the range of numbers in common use continues to increase.

Perhaps the most common cause of misunderstanding is confusing quantities and rates – as with energy and power, capital and income, distance and speed (familiar to mathematics teachers, though people are usually clearer on this one). Discussions of power generation and use – in the context of moves to mitigate global warming, for example – need to distinguish the rate of energy production in power stations, wind turbines, or solar panels (Gigawatts) from the household energy consumption that appears on a monthly bill (kilowatt-hours), or its national annual equivalent. The official SI unit of energy – the Joule – is largely neglected in this context (for what it's worth, 1 kilowatt hour is 3.6 Megajoules).

It is curious, and unfortunate, that experts are often sloppy about units – to the point of being misleading. Discussions about GDP and national debt are an egregious example. "This means the national debt will grow to 75% of GDP" is a typical statement in the media. But the national debt is a lump sum of money (or rather lack of it), whereas GDP is annual rate of expenditure of money, so "9 months GDP" is what they mean. When challenged, they say "Everyone knows what I mean". They don't. Nobody would say "The distance from Liverpool to Manchester is 50% of the motorway speed limit".

Back to our general point: units matter. They clarify meaning.

Acknowledgements

Figures 12.1.1 and 12.1.2 were created by the authors but were inspired by the work of the BA/MARS team at Harvard Graduate School of Education. Figure 12.1.3 is by the Math Assessment Project team and appears courtesy of Bell Burkhard Daro Shell Centre Trust. Figure 12.3.1 is adapted from a public domain image from Wikipedia (left) and an image by 'Woodstone' (right) licensed under the Creative Commons Attribution-Share Alike 4.0 International licence (https://creativecommons.org/licenses/by-sa/4.0/).

References

Black, P., Burkhardt, H., Daro, P., Jones, I., Lappan, G., Pead, D., & Stephens, M. (2012). High-stakes examinations to support policy. *Educational Designer*, *2*(5). Retrieved 16 October 2023. Retrieved from http://www.educationaldesigner.org/ed/volume2/issue5/article16

Burkhardt, H. (2009). On strategic design. *Educational Designer*, *1*(3). Retrieved 16 October 2023. Retrieved from http://www.educationaldesigner.org/ed/volume1/issue3/article9

Burkhardt, H., Fraser, R. E., & Ridgway, J. (1990). The dynamics of curriculum change. In I. Wirszup & R. Streit (Eds.), *Developments in school mathematics around the world* (Vol. 2, pp. 3–30). National Council of Teachers of Mathematics.

Drucker, P. F. (1995). *People and performance: The best of Peter Drucker on management*. Routledge.

Goodhart, C. A. E. (1975). Papers in Monetary Economics, Volume I, Reserve Bank of Australia.

Guadagnoli, T. (Ed.) (2000). *Advanced High School Assessment, Package 2 (Balanced Assessment for the Mathematics Curriculum)*. Dale Seymour Publications.

OECD. (2019). PISA 2018 Results (Volume II): Where *All Students Can Succeed*. OECD Publishing, Paris. https://doi.org/10.1787/b5fd1b8f-en

Links to useful material

To visit any of these links, scan this QR code or visit ltml.mathlit.org – append the link code to go directly to the entry: – for example, ltml.mathlit.org/12A

The original source links are given below for attribution purposes:

[12A] *Measure (mathematics)* – Wikipedia
https://en.wikipedia.org/wiki/Measure_(mathematics)

[12B] Example *'-ness' Tasks* – mathlit.org
https://ltml.mathlit.org/12B

[12C] *Designing a Measure: Correlation* – Mathematics Assessment Project
https://www.map.mathshell.org/lessons.php?unit=9410&collection=8

[12D] *The Long-run Differences between the CPI and RPI* – Office for Budget Responsibility
https://obr.uk/box/the-long-run-differences-between-the-cpi-and-rpi/

[12E] *Gini Coefficient* – Wikipedia
https://en.wikipedia.org/wiki/Gini_coefficient

[12F] *The SI Base Units* – National Physical Laboratory
https://www.npl.co.uk/si-units

13

MATHEMATICS FOR INFORMATION TECHNOLOGY

Chapter 10 discussed some of the many exciting ways in which Information Technology (IT) could support classroom teaching in ways particularly relevant to mathematical literacy. Sections 10.5 and 10.6 started to look at some examples of how the widespread use of technology was making new demands on mathematical literacy. In this chapter we approach the more complex question of what mathematics should students *know about* to help them engage in more depth with the technology that is now part of everyday life. Developments in computers, communications, and information processing have been profoundly changing our private lives and workplaces since the mid-twentieth century, and 'computer science' is now an important field of both pure and applied mathematics, yet this has had little impact on school mathematics curriculum. Here we aim to pick out a few examples where accessible and widely taught mathematics topics can be made relevant to the challenges of an increasingly digital world.

There is an entire field of 'discrete mathematics' that emerged in the mid-twentieth century and has grown in importance alongside, often in connection with, computer technology. Broadly, this is summarised as 'the mathematics of countable sets', in contrast to the more traditional 'continuous mathematics' of real numbers, algebra, infinitesimal calculus, and Euclidian geometry. In more practical terms, discrete mathematics forms a broad umbrella for topics such as the integers, formal logic, graph theory ('systems of nodes and edges' – not illustrative charts), and discrete calculus (the study of incremental changes as opposed to the infinitesimal changes of traditional calculus) as well as the fundamental theories underpinning computing and the study of algorithms. A modest example of the difference between discrete and continuous approaches is using a spreadsheet to model population growth via an

DOI: 10.4324/9781003303503-14

iterative formula (Figure 9.3.1) instead of formulating and solving a differential equation. Introducing 'discrete mathematics' to the curriculum was part of the 'new mathematics' courses of the 1960s, but it has not significantly displaced traditional mathematics. Yet the computers we know were partly conceived by mathematicians (most famously Alan Turing) and the close relationship between computer programming and mathematics are dramatic examples of how mathematics can 'change the world'.

In this chapter, we start by addressing how computers solve various types of problems and pick out a few examples of the mathematics behind computing, progressing from simple calculations using well-known formulae to the topical field of machine learning. This is followed by a look at just two examples from the vital field of computer security – choosing a strong password and a brief introduction to encryption, both of which are highly relevant to users of current technology.

We don't expect students to learn *how* to use and apply this mathematics in any detail. Instead, these sections highlight ideas that are valuable for a mathematically literate person to 'know about' in a technological world. Both 'know how' and 'know about' feature in the context-focused mathematical framework (see Figure 1.1.1). These ideas might form the basis of mathematical discussions or even essay projects. They are not central to most mathematics curricula, but some students will encounter them in various digital technology subjects, so there are opportunities for teachers to make interdisciplinary links. As we discussed in Chapter 1, this absence of 'know about' seems to be a particular feature of mathematics education – geography students, for example, aren't expected to reproduce the chemical equations for aluminium smelting in order to discuss why it is an important industry in regions with cheap renewable energy – they can 'know about' the energy demands without 'knowing how' they are evident from the equations. In contrast, mathematical tasks at school almost always ask 'know how' questions involving using a learned procedure, solving an equation, or proving a theorem. So this chapter aims to help readers 'know about' the mathematics underpinning important aspects of IT in action.

We have deliberately not addressed the teaching of programming here. Learning coding can be a valuable addition to the school curriculum – both as a practical vocational skill and as a source of deeper understanding of both technology and mathematics – but it is beyond the scope of this book. Realistically, acquiring any level of fluency in a particular programming language would take too much time to sneak in to most mathematics courses. We have included a couple of snippets of simplified code for the benefit of those who might have some experience of programming, but where possible we have turned to spreadsheets to illustrate points. We propose (see Section 3.3) that using a spreadsheet well as an important mathematical literacy skill for most people to have, to be developed partly within school mathematics.

13.1 Numerical calculations and accuracy

In the following sections, we look at some examples of how computers solve – or, sometimes, *fail* to solve – various problems. Some tasks seem straightforward to tackle using a computer, because they have been analysed mathematically and reduced to a formula. For example, the quadratic equation:

$$ax^2 + bx + c = 0$$

can be rearranged algebraically to give the familiar formula:

$$x = \frac{-b \pm \sqrt{b^2 - 4ac}}{2a}$$

This can easily be coded using a high-level programming language, a spreadsheet, or even some pocket calculators as something like Figure 13.1.1.

Most popular programming languages only differ on fine detail such as what the positive square root function is called (usually something like "sqrt"). Of course, the above is not a complete program – it doesn't input the values for *a, b, c* or output the results. Also, there are a couple of obvious special cases that the computer won't handle unless it is explicitly programmed to spot them: if the equation doesn't have any solutions (no real roots), then $b^2 - 4ac$ will be negative, and trying to take the square root using

```
a = 2
b = -6
c = 3
x1 = (-b + sqrt(b*b-4*a*c))/(2*a)
x2 = (-b - sqrt(b*b-4*a*c))/(2*a)
```

	A	B	C
1		Quadratic Equation ax² + bx + c = 0	
2	a =	2	
3	b =	-6	
4	c =	3	
5	discriminant =	12	← B3^2-4*B2*B4
6	x =	**2.366025404**	← (-B3+SQRT(B5))/(2*B2)
7	or	**0.633974596**	← (-B3-SQRT(B5))/(2*B2)
8			

FIGURE 13.1.1 Simple quadratic equation solution as code (top) and spreadsheet (bottom).

Source: Spreadsheet and code available on the website [13A].

Decimal		
100s	10s	1s
0	1	2
$10+2=12$		

Binary			
8s	4s	2s	1s
1	1	0	0
$8+4+0+0=12$			

FIGURE 13.1.2 The number twelve as decimal and binary.

the standard sqrt function will make the computer stop with an error, as will dividing by $2a$ if $a=0$, even though $bx+c=0$ has a solution.

In writing such a program, we take for granted that computers can 'just do' that sort of calculation. We won't go into the details of how, behind the scenes, everything ultimately gets reduced to simple logical operations on binary 0s ('off') and 1s ('on'). There are, however, some consequences that can derail your computer program and cause it to crash or give inaccurate results.

Computers represent *integers* in binary (base two). The number twelve (12_{10}) in base ten (decimal) becomes 1100_2 in base two (binary) (Figure 13.1.2).

What if we want to represent a fraction or mixed number in binary – say twelve and a quarter? The principle is the same as decimal – but instead of the place value of each digit after being one tenth of place value to the left, after the 'binary point', each column halves its place value, so $12.25_{10} = 1100.01_2$ (Figure 13.1.3).

Of course, as with the decimal system, some real numbers (e.g. fractions, square roots) will require an infinite number of digits to represent accurately in binary. Large base ten integers will need very large numbers of binary digits. Digital computers have space for a limited number of digits (called 'bits' in binary). Hence, real numbers are generally represented as 'floating point' numbers consisting of an integer part multiplied by an integer power of 2. This is the equivalent of scientific notation in decimal, where

$$0.000001234567 \left(=1.234567\times10^{-6} \text{ in proper standard form}\right)$$
$$=1234567\times10^{-12}$$

Decimal 12.25					
100s	10s	1s		$\frac{1}{10}$s	$\frac{1}{100}$s
0	1	2	.	2	5
$12+\frac{2}{10}+\frac{5}{100}=12\frac{1}{4}$					

Binary 1100.01						
8s	4s	2s	1s		$\frac{1}{2}$s	$\frac{1}{4}$s
1	1	0	0	.	0	1
$8+4+\frac{0}{2}+\frac{1}{4}=12\frac{1}{4}$						

FIGURE 13.1.3 The number twelve and a quarter as decimal and binary.

So the 1234567 part (the mantissa or significand) could be stored as a seven-digit integer and the exponent (–19) as a two-digit integer and it could all be stored in ten digits. In binary it needs more space (bits).

$$\begin{aligned} 0.001234567_{10} &\approx 0.000000000101000011101000100110010_2 \\ &= 1.0100001110100010011001 0_2 \times 2_{10}^{-\text{ten}} \\ &= 1.01000011101000100110010_2 \times 10_2^{-1010\ (\text{binary})} \end{aligned}$$

However, note the "≈" in the first line – that is only an *approximation* because most fractions that can be written precisely as decimal fractions are recurring binary fractions. For example:

$$0.1_{10} = 0.00011001100110011001100110011001100110011\ldots_2$$

> **ACTIVITY**
>
> As in base ten, all fractions either terminate or repeat in binary. Investigate which fractions terminate in binary and which repeat.

So, if you enter 0.1_{10} (precisely 1/10), then behind the scenes it is getting stored as a recurring binary fraction truncated to (usually) 23 or 52 'binary places' and it is not *quite* equal to 0.1_{10}. That shouldn't be surprising – many real numbers can't be expressed precisely as decimal fractions either – but in a computer program or a spreadsheet you can be fooled by entering what looks like a precise value in decimal resulting in 'rounding errors' that become significant over multiple operations. Figure 13.1.4 shows this happening in a spreadsheet – after multiple additions of 0.01, the result is very slightly wrong. This looks innocuous (you'll have to increase the number of decimal places to the maximum to see it) but the result is that the value never reaches *exactly* 2346. This sort of thing can cause problems for programmers who assume *exact* results. The simple program in Figure 13.1.5 starts with f = 2345, repeatedly adds 0.01 and is told to stop when f = 2346. We intend it to reach 2346 after 100 iterations and then stop. But this program will likely fail and keep running indefinitely because, like the spreadsheet, the closest f gets to 2346 is 2346.00000000002 which is *not* equal to 2346. So, when programming with floating point numbers, it is *very* important to avoid checking that two numbers are exactly equal, or to rely on the result of subtracting two closely equal numbers.

Let's say you fix the program in Figure 13.1.4 to stop when f ≥ 2346, so it does actually finish – it then prints $2.1827872842550278 \times 10^{-11}$ instead of 0. That's a very small error *but* your programming language was meant to be accurate to 15 significant digits. This demonstrates how rounding errors accumulate in complex programs.

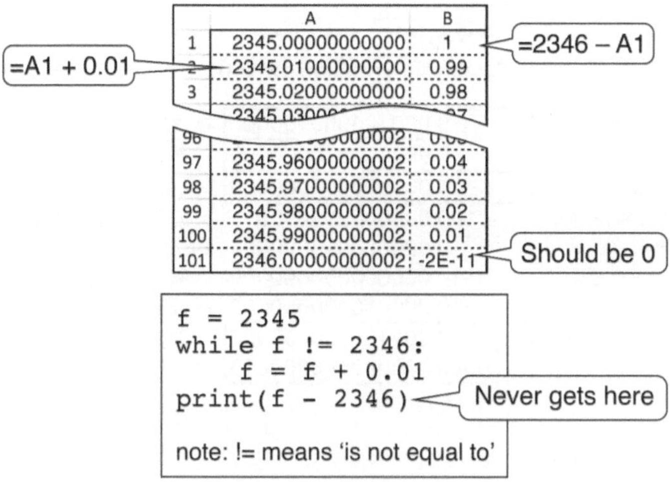

FIGURE 13.1.4 Rounding errors in spreadsheets (top) and code (bottom).

Source: Spreadsheet and code available on the website [13B].

The limited precision of floating-point numbers can trip up programs like our quadratic equation calculator if we want high accuracy. Say we wanted the computer to solve:

$$\frac{x^2}{10\,000}+\frac{999999999x}{100\,000\,000}-\frac{1}{1000}=0$$

$$\left(\textit{Note}\text{: roots } x=\frac{1}{10\,000}\text{ and } x=-100\,000\right)$$

A computer algebra system will give these two roots precisely, because it solves the equation analytically, by rearranging and factorising, and will keep the fractions *as* fractions where possible. However, our spreadsheet (see Figure 13.1.5)

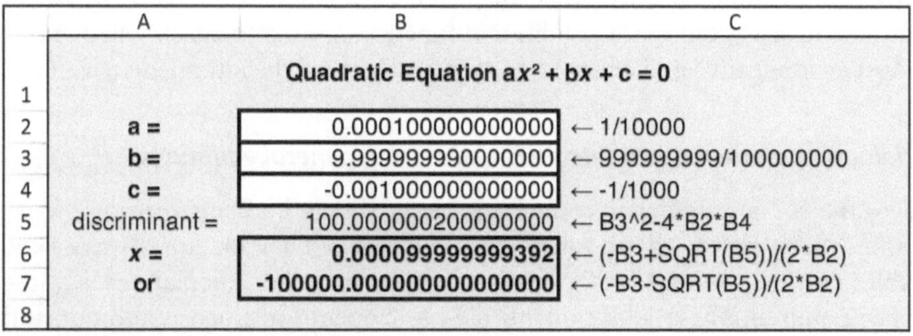

	A	B	C
1		Quadratic Equation $ax^2 + bx + c = 0$	
2	a =	0.000100000000000	← 1/10000
3	b =	9.999999990000000	← 999999999/100000000
4	c =	-0.001000000000000	← -1/1000
5	discriminant =	100.000000200000000	← B3^2-4*B2*B4
6	x =	**0.000099999999392**	← (-B3+SQRT(B5))/(2*B2)
7	or	**-100000.000000000000000**	← (-B3-SQRT(B5))/(2*B2)
8			

FIGURE 13.1.5 Spreadsheet quadratic solution showing rounding error.

Source: Spreadsheet available online [13C].

	A	B	C
1		**Quadratic Equation** $ax^2 + bx + c = 0$	
2	a =	0.000100000000000	← 1/10000
3	b =	9.999999990000000	← 999999999/100000000
4	c =	-0.001000000000000	← -1/1000
5	discriminant =	100.000000200000000	← B3^2-4*B2*B4
6	if b < 0 q =	0.000000010000000	← -0.5 * (B3 - SQRT(B5))
7	if b ≥ 0 q =	-10.000000000000000	← -0.5 * (B3 + SQRT(B5))
8	q =	-10.000000000000000	← IF(B3 < 0, B6,B7)
9	**x =**	**0.000100000000000**	← B4/B8
10	**or**	**-100000.000000000000000**	← B8/B2
11			

FIGURE 13.1.6 A more accurate spreadsheet quadratic solution.

Source: Spreadsheet available online [13C].

gives the roots as $9.99999993922529 \times 10^{-5}$ and $-100\ 000$ – again not the 15 significant digit precision we were expecting. To make it worse, we could have *guessed* $x \approx 0$ if we wanted an approximation. The problem is, in our quadratic formula, if ac is much smaller than b, then $\sqrt{b^2 - 4ac}$ is going to be very close to b and so calculating one of the roots will involve subtracting two very similar numbers, and we have seen from Figure 13.1.4 that this can leave a result which is dominated by rounding errors. The moral is that you can't always just translate a mathematical formula into your favourite programming language or spreadsheet and expect precise results. Instead, it is sometimes necessary to rewrite the formula to avoid rounding errors in intermediate results – see Press et al. (1986, p. 145). A better – but harder to follow – spreadsheet to solve our equation is shown in Figure 13.1.6. This appears to just be an unnecessarily round-about way of calculating the same thing, but which avoids *subtracting* $\sqrt{b^2 - 4ac}$ from b where the two are very close and the difference would be affected by rounding errors. *Note:* Row 8 in the spreadsheet contains a conditional function needed to check whether b is positive or negative and choose one of two values for the intermediate value q.

Evaluating mathematical functions and solving general equations

We take for granted that computers know things like square roots, logarithms, and trigonometric functions – but how? They do not always have tables of values stored inside but use inbuilt rules that calculate each value when required. These rules can only use basic operations such as addition and subtraction, multiplication and division. The mathematics behind this is beyond any expectations of mathematical literacy, and the algorithms actually used are heavily optimised for speed. However, some of the basic concepts feature in more advanced high school and college mathematics courses, and, when they do arise, it is worth making the connection to computing as an important practical application.

Taylor's series (and similar series expansions) can turn functions into an infinite series of terms, for example:

$$\sin(x) = x - \frac{x^3}{3!} + \frac{x^5}{5!} - \frac{x^7}{7!} + \cdots$$

The first terms of this series could be used to estimate the sin() function using just multiplication, addition, subtraction, and division – the more terms, the greater the accuracy. Computers probably do not use this exact method, for one thing it is probably too slow, but the idea helps understand how it is possible for computers to 'know' these functions. An example spreadsheet calculation using this expansion is shown in Figure 13.1.7. The value of x is entered in cell E1. Column A gives the position numbers of each of the (first seven) terms. Column B calculates the odd numbers for the denominators, and Columns C and D enable calculation of each term. In column E, the values of the terms are successively added. The example shows a good estimate for sin $\pi/6 = 0.5$ is given with just a few iterations.

Some other mathematics topics from college and higher level secondary school classes are related to algorithms that are built into computers (in an optimised form):

- *Newton's method* to improve an estimate of the solution to an equation. Computers often use this and similar methods to calculate square roots.
- *Trapezoid rule* for estimating integrals (likewise, estimating the first derivative from two closely spaced points on a curve). The easiest way for computers to do calculus.

	A	B	C	D	E	F
1				x=	**0.52359877560**	← PI()/6
2					(in radians)	
3		**N**	**Sign**	**N!**	**Sum**	
4	1	1	1	1	0.52359877560	← C4*POWER(x,B4)/D4
5	2	3	-1	6	0.49967417939	← E4+C5*POWER(x,B5)/D5
6	3	5	1	120	0.50000213259	← E5+C6*POWER(x,B6)/D6
7	4	7	-1	5040	0.49999999187	← E6+C7*POWER(x,B7)/D7
8	5	9	1	362880	0.50000000002	← E7+C8*POWER(x,B8)/D8
9	6	11	-1	39916800	0.50000000000	← E8+C9*POWER(x,B9)/D9
10	7	13	1	6227020800	0.50000000000	← E9+C10*POWER(x,B10)/D10

FIGURE 13.1.7 Taylors series expansion of sin(x) using a spreadsheet.

Source: Spreadsheet available online [13D].

- *Method of differences* can make summing series far more efficient. This is significant to the pre-history of computing, as Charles Babbage's Difference Engine was a mechanical calculator intended to use this method to calculate scientific tables. The Difference Engine was never completed, but subsequent mechanical devices used the same theory.

13.2 Storing and organising data

Sorting

One of the most widespread uses of computers in business and commerce is storing and organising large databases. A database needs to be easy to update, easy to keep consistent, efficient to search, and capable of producing 'reports' which summarise or reorganise the data in useful ways.

A common task in data storage is sorting lists into numeric, alphabetic, or other order – a much-studied subject in computer science on which whole books have been written (famously Knuth, 1973, weighing in at 723 pages – plus fold-out!). If you are sorting millions of items, having the most efficient algorithm can be vital.

How would you instruct a computer to sort items? You could experiment with 'rules' for sorting using, say, some letter tiles from a word game:

T K P E Y O X S A

Looking at those, you would immediately start thinking "oh, X and Y go at the end, the A on the front, then the E" and the job would be half done. But what if you (like a computer) didn't have that sort of human insight, and could only pick two letters from the list and ask which one should come before the other? One algorithm – called a 'bubble sort' – is particularly easy to understand and illustrates the sort of logic that can be simply programmed into a computer.

1 Work through the list from left to right, one at a time and compare each letter with the following letter.
2 If the two letters are in the wrong order, swap them.
3 Keep going until you've checked the last two letters.
4 If you had to make any swaps, go back to step 1.
5 Keep going until you make a pass through the list without making any swaps – your list is now sorted! No cheating – with these rules the computer wouldn't notice that the job is finished until it had gone through the list comparing every pair of letters that one last time.

It is worth noting that this is a horribly inefficient process, as you'll see if you try to follow it by hand. In the example you need 8 passes to get the

A from one end of the list to the other, meaning 64 pairs of letters to compare. Even without such a 'worst case', this is what is known as an $O(n^2)$ or "order n^2" algorithm, because the number of steps tends to go up with the *square* of the number of elements in the list. This is not a problem if you're sorting a handful of items, but it is a problem when there are millions of items. You can probably work out a few ways to make the sort more efficient. Today, most programming languages have more sophisticated sort routines built-in and ready to use. These are too complex to describe to a human but can sort a list using the order of $n \log n$ operations. Most programmers wanting a simple sort would use an existing tool – the above examples are intended as an illustration of how computers tackle this sort of straightforward but laborious tasks, and to raise awareness of the attention that goes into doing apparently straightforward tasks efficiently.

Efficient searching, 'keys', and relational databases

One benefit of sorting a list of items, as discussed in the previous section, is that it is then easy to search, say, a list of names using an efficient "binary search" without having to check every single list item.

1 Look at the value in the middle of the list.
2 Does it come before or after your target name in alphabetic order?
3 If before, discard the top half of the list. If before, discard the bottom half of the list.
4 Repeat until you find the target.

In a large database, however, the entries are often not in any particular order, and there may be several different things you may want to search for. You can't resort whole database for every different search. The solution is to produce one or more 'indices' or 'keys'. Figure 13.2.1 gives a trivial example

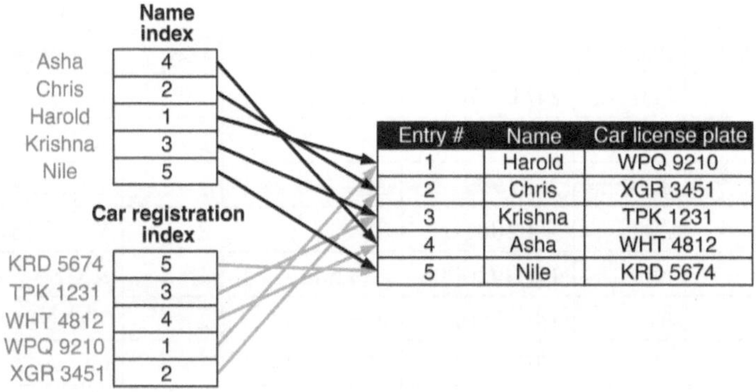

FIGURE 13.2.1 Database of car licence plates with two indices.

Employee ID	Name	Date of Birth	Room #	Desk #	Car licence plate
1	Harold	21/3/1987	3A	1	WPQ 9210
2	Chris	14/6/2002	3A	2	XGR 3451
3	Krishna	3/7/1999	3B	1	TPK 1231
4	Asha	9/8/2001	3B	2	WHT 4812
5	Nile	17/10/2000	4	1	KRD 5674

FIGURE 13.2.2 A flat file database.

of a database of car license plates with one index to help find entries by name and another to search by license number. Each index is just a list of numbers that point to entries in the main database, sorted in alphabetic order of owners' names or car license numbers, respectively, which can be used to do a rapid binary search as described above.

A development of this idea – the *relational database* – is used in most large business databases today. There is a sophisticated mathematical theory behind relational databases, but one fundamental idea is that every important piece of information should only be stored in one place, making it easy to update and hard to add multiple, conflicting entries. Imagine we are trying to extend the database in Figure 13.2.1 to keep track of employees, their personal info, their cars (for keeping track of parking), and where they sit. We could extend it as in Figure 13.2.2. This is fine, until 'life happens'. Chris has a second car that they want to park sometimes (and, by the way, their name is spelled "Kris") and then Shana joins the company part time working Monday–Wednesday sharing Nile's desk (since he only comes in on Thursdays and Fridays) and room 3A gets renamed "The Burkhardt Suite". So pretty soon you end up with something like Figure 13.2.3, with duplicated entries, contradictions, and non-standard annotations.

Employee ID	Name	Date of Birth	Room #	Desk #	Car licence plate
1	Harold	21/3/1987	Burkhardt	1	WPQ 9210
2	Chris	14/6/2002	3A	2	XGR 3451
3	Krishna	3/7/1999	3B	1	TPK 1231
4	Asha	9/8/2001	3B	2	WHT 4812
5	Nile	17/10/2000	4 (Thurs–Fri)	1	KRD 5674
2	Kris	14/6/2002	Burkhardt	2	TYB 9876
6	Shana	12/2/1998	4 (Mon–Wed)	1	none

FIGURE 13.2.3 Flat file database 6 months later.

Of course, if we were only keeping track of half a dozen employees, we would be massively overthinking this. In reality, the database would be for a firm with hundreds of employees and probably several people responsible for updating the database, so the mistakes wouldn't be so obvious, and (say) trying to trace an improperly parked car or discover which desks were free on particular days would become error prone. What is needed is a database that makes it hard to enter invalid data, easy to update, and is easier to restructure when new requirements (such as desk sharing) arise. If we look at our example, we can make some rules about the data.

- Every person has exactly one name, date of birth, and can be allocated a unique employee ID which shouldn't need to change.
- Not all employees have cars but some have more than one.
- Each car is owned by exactly one employee (we can worry about car shares later – the car will still have one owner).
- Each car has a unique licence plate number and no two cars should have the same number.
- Employees can share desks, but only one employee can use a desk on a particular day.
- Each desk is in a room, and has a number that is only unique within that room.
- Each room has a name.

Now we apply the principle that each bit of important data should be stored only once. So, we could organise our (original) data as in Figure 13.2.4. The data is now spread across several tables. "Employees" contains *just* the

Table: Employees

Employee ID	Name	Date of Birth
1	Harold	21/3/1987
2	Chris	14/6/2002
3	Krishna	3/7/1999
4	Asha	9/8/2001
5	Nile	17/10/2000

Table: Cars

License Plate	Owner's Employee ID
WPQ 9210	1
XGR 3451	2
TPK 1231	3
WHT 4812	4
KRD 5674	5

Table: Seating

Desk ID	Employee ID
1	1
2	2
3	3
4	4
5	5

Table: Desks

Desk ID	Room ID
1	1
2	1
3	2
4	2
5	3

Table: Rooms

Room ID	Room Name
1	Room 3A
2	Room 3B
3	Room 4

FIGURE 13.2.4 Relational database.

personal details of each employee. "Cars" just contains car licence plates and the ID of the employee they belong to. So, you could look up a car licence plate, find the employee number and then look up the employee's ID in "Employees".

That seems overcomplicated, until you add Kris' second car. All you need to do is add the license plate number and Kris' employee ID to the 'Cars' table and you're done. Looking up either license plate in "Cars" will give you Kris' ID, and looking up Kris' ID will find *both* cars belonging to Kris. Meanwhile, there's still only one "Employees" entry for Kris, so we don't end up with two copies of their personal information which could get out of step. If Kris gets rid of the second car, you can just delete it from "Cars" without breaking anything. So we have a "one to many relationship" – one employee can have many cars.

The shaded column in each table of Figure 13.2.4, which provides a unique, unchanging ID for each entry in the table and which can be referenced by other tables, is called the 'Primary Key'. Database software can ensure the integrity of keys by, for example, refusing to add a new row with a duplicated primary key, or deleting a row with a key that is referenced by another table. (In practice, a database designer would probably give each table an extra column of guaranteed unique values as a primary key rather than relying on externally supplied values like Employee IDs and car licenses to be unique.)

The "Rooms" table gives each room a name. "Desks" assigns each desk an ID and links it to a room and "Seating" assigns each employee to a desk. So, when you rename Room 3A, it only takes one change to the row for Room ID 1 to change the name for everyone.

That looks really overcomplicated until Shana wants to share a desk with Nile. Then it means we could simply add a line to "Seating" to link Shana's employee ID to desk ID 5 (and accept that each desk could have more than one user). If we wanted to keep track of who used what desk on what day, you could add a day-of-week column to the "Seating" table. For each employee ID, there would be one entry for each day of the week that they used a particular desk.

The specifications of these tables would be input to a database management system. The details would include rules about which columns referenced other tables, which columns (or groups of columns) always had to contain values that were unique within the table, which could be left blank, etc. The database management system would refuse to make changes that broke the rules.

All this is far more complicated than just typing the information into a spreadsheet. In fact, a database like this would usually require a programmer to design forms and 'helper code' that made it easy for the users of the database to enter and update information without worrying how it was split up between tables. However, a well-designed database like this will help ensure that the data is always consistent and any updates are properly applied. Spreadsheets are fine until they grow over the years and become unmanageable and full of errors. Relational databases, as described here, are not the only type of database, but they have been a mainstay of commercial computing since the late 1980s and most of the world's data management systems in government, healthcare, and business run on them.

13.3 Heuristics and 'calculated guesses'

You would usually assume that a computer program was designed to reliably produce the "correct" result. True, a program based on a mathematical formula will have limitations on the range of parameters it can accept and the mathematical precision of the result, while a sorting program could be overwhelmed with too many items, or fed with characters that it couldn't deal with, but fully developed versions of those algorithms would detect and warn of such problems. In theory, within those constraints, and subject to any assumptions and simplifications made in formulating the model, the results will be provably correct.

This is not always the case. Not all problems can be reduced to a mathematical formula or a simple, reliable set of steps. Other methods are needed to tackle problems that are too difficult to fully analyse or where computing the exact result would take too long or indeed may be impossible. In fact, Turing and others proved mathematically that there are some problems for which it is theoretically impossible to precisely compute the solution. Inconveniently, the proof does not say which problems are like that.

For problems which cannot be analysed in practice, it is often possible to use a *heuristic* method to produce a "good enough" result by trial and error or by applying rules of thumb. These rules do not fully describe the problem, and there is no guarantee that the result will absolutely always be correct. But it is all we have. Although the first example below is a game with no serious real-world use, it illustrates a distinction which has important implications for many real-world applications of computers which rely on such heuristic methods and can never guarantee a correct answer.

Playing games

Take a game like noughts and crosses (also called tic-tac-toe)– how could you program a computer to play it? Figure 13.3.1 shows the eight steps of a game

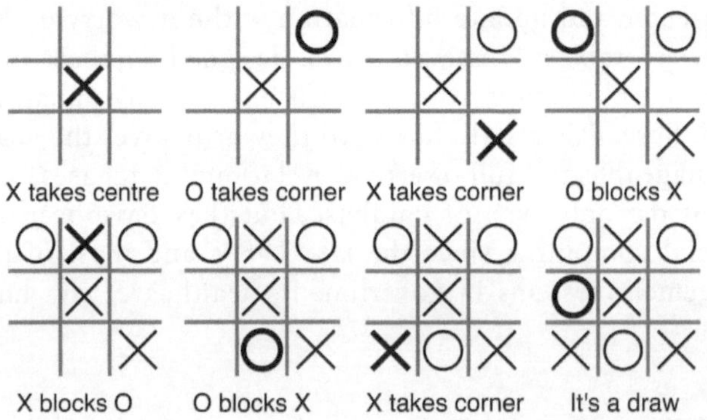

FIGURE 13.3.1 A typical game of noughts and crosses (tic-tac-toe). Both players are following the rules described in the text.

that ends in a draw. If you've played the game more than a few times, you've probably learned a strategy something like:

1 If your opponent can win on their next move by completing a row or diagonal, block them!
2 Otherwise, if the centre square is available, take that.
3 If not, and a corner square is available, take that.
4 Otherwise, pick any free square.

That would form a set of "heuristics" that could easily be turned into a computer program. It would play a good enough game to beat someone with no experience, but it is not guaranteed to win or draw every time. In fact, noughts and crosses is sufficiently simple that it is a "solved game". It is quite practical for a modern computer (or even a good high school student) to analyse *all* the possible moves from a given point, pick the best outcome, and play a 'perfect' game, guaranteeing to, at least, draw. However, the four rules above – plus, maybe remembering a few '1 move to win' positions – are probably closer to how a human would play the game.

ACTIVITY

How many possible games of noughts and crosses are there?

(An upper limit to the number of possible games is 9 possible positions for the first move, 8 for the second and so on, giving $9 \times 8 \times 7 \times 6 \ldots \times 1 = 9! = 362880$. Deeper thought shows that is an overestimate as, for example, some games don't continue until the board is full. And appreciating symmetry shows that there are only three really different first moves: centre, corner, or side.)

For a more complicated game, such as chess, there are many more possible games (about 5×10^{44} board positions estimated to give about 10^{120} possible games, versus ~10^5 for tic-tac-toe) so a complete solution (an algorithm guaranteed to win or draw every game) is still beyond the power of current computers. Instead, they combine looking a few moves ahead and applying 'heuristic' rules or looking for correspondences with famous games. No chess-playing program is mathematically guaranteed to win every game. Yet, the top chess programs have now beaten grandmasters – and even years before that had happened, chess-playing programs could usually thrash a casual player. These are still successful commercial products.

Playing games may not seem like a 'practical' way of using computers, but games provide insight into how computers solve other problems. Enumerating possible outcomes or factors and 'scoring' the results (in the noughts and crosses, the scores are win, draw, and loss), while looking for shortcuts to reduce the number of paths that need to be explored is a widely used approach in computer 'intelligence'. These methods apply from medical assessments to deciding whether you get a bank loan. The take-home message for mathematical literacy is the important distinction between a computer *solving* a problem in the rigorous mathematical sense versus being able to tackle a "real" problem as well as, or better than, a typical human by making what is essentially a 'calculated guess'.

Monte Carlo methods

Another method of finding approximate solutions problems that may be too complex to fully analyse mathematically is to test it for a random sample of values and use probability to estimate the result. These are called Monte Carlo methods, named after the municipality in France that is famous for having many casinos. For example, Figure 13.3.2 shows how a Monte Carlo method can be used to estimate the area of an irregularly shaped oil spill by picking 200 points at random on the image and calculating the proportion that are inside the spill. This gives an estimate of the probability a randomly chosen point will be inside the oil spill which in turn estimates the ratio of area of spill to the area of map.

This example comes from the Australian Academy of Science's *reSolve: Maths by Inquiry* materials [13E] and is suitable for classroom use – in the lesson, students compare the accuracy of the Monte Carlo method with other geometrical ways of estimating the area. Other methods – such as splitting the shape into triangles – struggle with an irregular shape such as the slick shown. Even sampling a regular array of points risks missing some features which fall between the grid. Randomly chosen points might also miss important features, but if a lot of points are chosen, the chances of that are small.

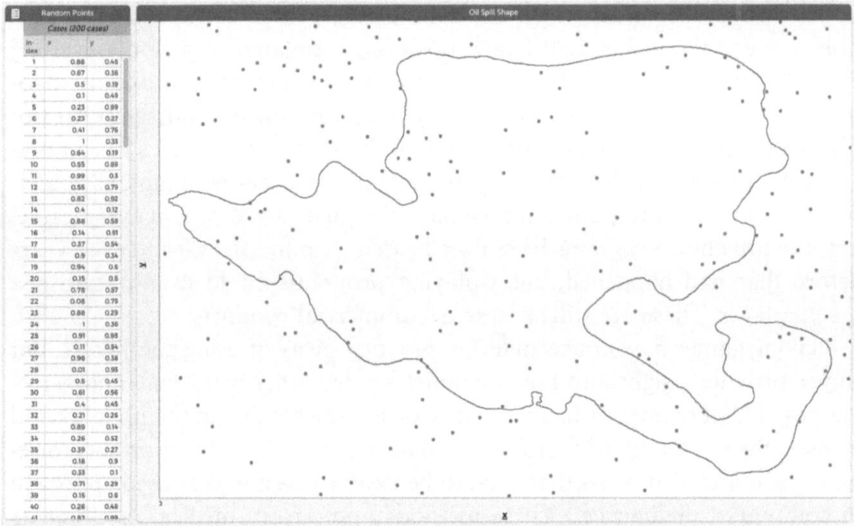

FIGURE 13.3.2 Finding the area of an oil spill by a Monte Carlo method.

Source: From *reSolve: Maths by Inquiry* & *CODAP* by The Concord Consortium [13E].

As with any classroom-friendly activity, the example is sufficiently simple that a modern computer could solve the problem by "brute force" simply counting each pixel in the image – but in more complex real-world applications, or where calculations have to be made rapidly (checking products rolling off a production line, for example), such methods can greatly reduce the number of data points that have to be tested.

13.4 Neural networks and machine learning

One way of solving a problem that cannot be completely analysed using mathematics is to use a "neural network". These are widely used in modern artificial intelligence and machine learning techniques. Neural networks mimic the way that biological brains are believed to work (greatly simplified!). Rather than being built to solve a problem using mathematics and logic, they are "trained" to produce the correct result using a set of known examples.

Figure 13.4.1 shows a diagram of a simple neural network. Each of the circles in the diagram represents a "neuron" with multiple input values and a single output. The network starts on the left with the initial input values, and moves right working through the various layers (columns) of neurons to the output.

Each input to a neuron has an associated 'weight' (illustrated here by the thickness of the lines). The neuron works by multiplying each input value

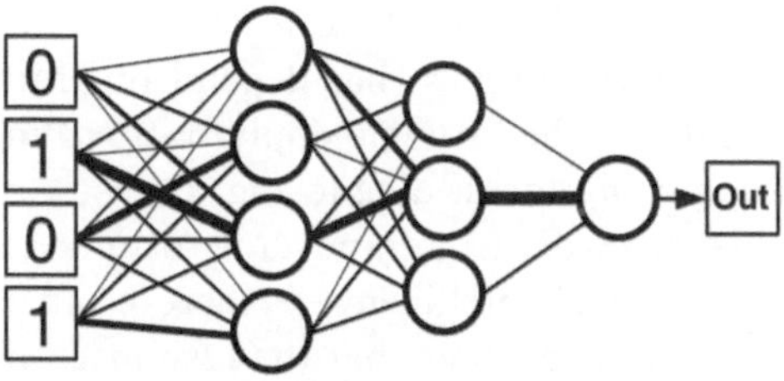

FIGURE 13.4.1 Simplified neural network.

by its weight, adding together the results and then applying an "activation function" to calculate the output value (this function could be as simple as "if total is greater than 0 then output 'cat', else output 'dog'"). This output is then passed on from that neuron to all of the neurons in the next 'layer' that it is connected to. In Figure 13.4.1 all the neurons are connected to all in the next layer, but this is not usually the case.

This network takes four numbers (or a four-digit number) as an input and produces a single output. You wouldn't normally use a network to do arithmetic, but the inputs could be (say) a person's height, weight, waist measurement, and shoe size gleaned from their online shopping habits, and the required output could be their approximate age, propensity for a certain illness, or the probability of them being interested in buying a smartwatch. A network of nine inputs could be the state of a noughts and crosses board, and the output would be the next move. A larger network, maybe with hundreds of inputs and more layers of neurons, might take the pixels of an image and output a code giving which letter of the alphabet it 'saw'. An image recognition system might reduce the image of, say, a face to a single, unique number representing that face, regardless of the exact lighting, angle, and framing in the image.

In any case, the network starts off by producing nonsensical output and must be 'trained' on a set of inputs with known answers. This is done by inputting an example, comparing the output to the correct result, and then tweaking the weights of each neuron's input accordingly – and repeating this many times with a large set of examples until the results have the required degree of accuracy. The process can usually be done automatically given a database of examples and a carefully designed training algorithm. Once trained, not only should the network produce the correct answer for any of the example inputs, but the goal is that it should work for inputs that it *hasn't* been trained on. This has to be verified by testing. Note that words like 'training' or 'learning' are widely used for this process but shouldn't be taken too literally. The design of the training algorithm is critical and its aim is to efficiently search for the optimum values of the weights.

The rapidly growing field of machine learning often uses neural networks, as well as other techniques that depend on training a computer on a bank of known data. The problem could be anything from targeting advertisements, recognising faces, diagnosing diseases, or making a self-driving car recognise hazards. It is important to understand that – even when advertised as artificial intelligence – there is no conscious understanding involved and no guarantee that the responses are correct. While there are ways of producing data to justify why a machine learning model produced a particular – possibly incorrect – response, this is usually much harder than for a more traditional model, where you might be able to point to a line of code, or an assumption made by the modeller to explain the result. Recently, my photo software automatically classified a photo I took of a white fungus as a cat. It would probably be impossible to describe to a human the features of the photo's pixels which caused the algorithm to select 'cat'. So it is especially important that users of such systems are alert for implausible results and also understand the mathematics of false positives and negatives and how they can sometimes lead to counterintuitive situations. This is discussed in more detail, in the context of medical testing (in which machine learning solutions are already playing a role), in Section 4.5.

Systematic errors have already been found to arise in machine learning systems due to a bias in the data used for training. For example, a machine learning system might be used to identify good applicants for places at a university, with the hope that it will be able to avoid the biases that human selectors implicitly apply. But if it is trained on the dataset of previously successful applicants, the biases of previous selectors will be built in. Such problems are exacerbated by the difficulty of justifying *why* a machine learning system made a particular decision – especially in lay terms – when there is no single rule that the developer can point to or simply correct.

13.5 Picking a password

Today, many websites, phone applications, video and music services, and other tech activities require you to create an account and choose a password. You soon end up with dozens of passwords to remember. The easiest thing is to use the same password for everything, but that is risky since if an attacker manages to guess or steal your password for one service, they have access to all of your other accounts. Even if you don't choose an easy-to-guess password, it only takes one incompetently designed website, a 'shoulder surfer' or being fooled by a 'phishing' email for it to leak out. So really, you should choose a unique, clever password for every service you use and remember them all.

How do you choose a good password? There are several requirements.

1 The password has to be hard for a human to guess – so, no mother's maiden names or old chestnuts like "password", "secret", "12345678" or "swordfish". You can always find out the ten most commonly used passwords with a web search.
2 It has to be hard for a *computer* to work out by brute force (just testing everything), which is a lot to ask given the speed of modern computers. Moreover, a hacker using a computer to break a password can be smart by focusing their search first on common names or words, and variations such as having 'i's and 'o's turned to ones and zeros.
3 You have to be able to *remember it* and type it quickly. Something like "Qz$t17!pTT" would be a good, strong password that is highly unlikely to be guessed, but not easy to type or remember. This means it would likely end up written down.

One mathematical measure of password strength, taken from information theory, is *information entropy*. For a password this is calculated as

$$\log_2(\textit{number of possible passwords})$$

with the unit "bits of entropy". This measure is useful for comparing rules for valid passwords, or methods of choosing memorable passwords. It is those rules that determine the number of possible passwords to use in the formula. Why use this measure? Say it was possible to use a completely random 64 digit (bit) binary integer as a password: the number of possible passwords would be 2^{64} – so it would take up to 2^{64} systematic guesses to be sure of finding the password. Pretty secure. The entropy of such a password would be $\log_2(2^{64}) = 64$ bits. So the entropy for different password schemes tells us the number of random bits that would provide equivalent security. Another way of interpreting it is that that chance of guessing a password with 64 bits of entropy on the first try is the same as getting 64 successive heads by tossing a fair coin.

In reality, your choice of password will be more constrained – long binary numbers would be tedious to type and hard to remember. A totally random 8 letter password made from the 26 letters of the alphabet would be one of 26^8 possible passwords and so have an entropy of $\log_2(26^8) = 37.60$ bits. Not as good as our random 64-bit integer. But eight truly random letters are still hard to remember. An eight-letter *English word* would be easier to remember – but also easier to guess – since a computerised password guesser will doubtless run through the dictionary before it starts trying every random permutation. There are about 80,000 eight-letter English words, so the entropy is only $\log_2(80000)$ or about 16 bits, and of course, people only know a fraction of those words, reducing the entropy even further. Of course, both of those calculations assume that the guesser knows your password is eight letters long and that it is an English word.

Note that we are making the reasonable assumption that a hacker will run an *organised* search, trying dictionary words and well-known password schemes (pa55w0rd, password42, #password etc.) before resorting to a 'brute force' search along the lines of AAAAAAAA, AAAAAAAB... ZZZZZZZZ. Entropy is more of a rule of thumb guide rather than a precise measure of 'security level'.

ACTIVITY

What would be the entropy of a password selected by picking a lowercase English word between 5 and 10 letters long? How does this compare with a random lowercase password between 5 and 10 letters long? How would allowing any mixture of upper- and lowercase letters change these entropies?

Number of letters	*Approximate number of English words in the dictionary*
5	158 000
6	20 000
8	80 000
9	41 000
10	35 000

Studying password choice and strength is an interesting mix of mathematics and psychology. An important finding is that the usual password rules can actually make passwords more predictable. So, requiring a ten-character password to include at least one digit 0–9 in addition to the 26 letters of the alphabet would appear to increase its entropy from $\log_2(26^{10}) = 47$ bits to approximately $\log_2(36^{10}) = 52$ bits, assuming that the password consists of ten *purely random* characters. A human being is likely to start with one of a few thousand common five- to eight-letter words (only 10–12 bits of entropy) and then use one of a handful of strategies (change the 'I's to 1's, the 'o's to zeroes, append a digit). So, instead of just checking for the 2000 most common words, a hacker needs to check for, maybe eight variations of each of those words. This sounds significant but only adds 3 bits of entropy.

$$\log_2(2000 \times 8) = \log_2(2000) + \log_2(8) = \log_2(2000) + 3$$

Using *two* common words actually increases the entropy from $\log_2(2000)$ to $\log_2(2000 \times 2000)$ – that is, it *doubles* the entropy. A better strategy for a memorable password might be to stick to the letters a–z and string together several common words to make a nonsensical – but memorable – phrase. This was famously (at least within IT culture) illustrated by Randall Munroe in the XKCD comic strip [13F] who illustrated how a passphrase like 'correcthorsebatterystaple' was both more secure and easier to memorise than a single word mangled to something like 'Tr0ub4dor&3' to meet typical password rules.

ACTIVITY (FOLLOWING ON)

Study the cartoon in Figure 13.5.1. What assumptions have been made to estimate the entropy of both "styles" of password?

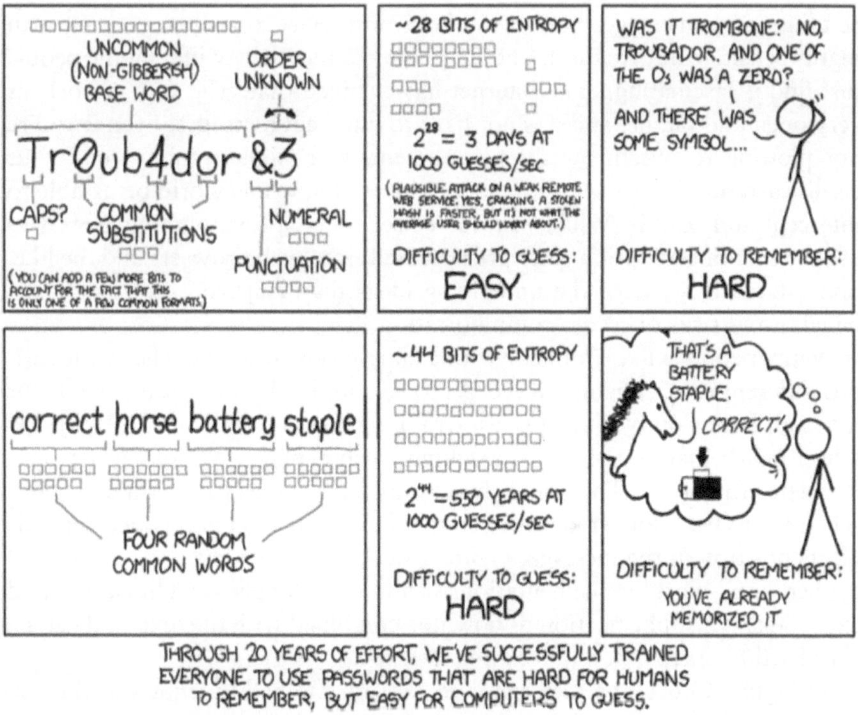

FIGURE 13.5.1 Choosing a password.

Source: From xkcd.com [13F].

Another solution to the "strong but memorable password" conundrum is to use a password manager application. The software equivalent of a key safe, these applications will generate unique, hard-to-guess (but equally hard to remember) random passwords for every website or service you use, and store them securely, so you only have to memorise one master password. That is the one to unlock the password manager. You can even use fingerprint or face recognition to quickly unlock it without typing a password. Of course, this means you have a file containing *all* of your passwords sitting on your computer and, in fact, you'll often want to keep it *online* so you can use it from your mobile devices and appliances. This would be disastrous in the wrong hands. There is a nice risk/benefit balancing issue there. Does the risk of having your passwords stored in a file outweigh the benefit of having strong, unique passwords for all your online activity? Of course, it would help greatly if the password file could be *so* strongly encrypted that it was useless to anybody without your master password.

13.6 Cryptography

Sending secret messages has always been of interest to people working with highly sensitive information, but not so much in everyday life. (Some people just find it fascinating.) The internet has changed that. The way it works is very open and public, and it is not hard to intercept messages. Even if you're not plotting to overthrow the government, you might want to send your credit card number to an online shop without the whole world being able to intercept and read it. While you don't need to be a cryptography expert to use the sort of encryption software now built into web browsers and the like, understanding some of the underlying ideas may help you understand the benefits and risks of secure communication.

Many people will be familiar with a simple substitution cypher with various different rules– change all the 'A's to 'C', all the 'B's to 'Q's, and so on, or move all letters on by three places, wrapping round from Z to A. A popular school mathematics activity is 'cracking' such cyphers by looking for clues like letter frequency. This shows how weak such cyphers are. A much better scheme is to use a longer keyword so that how each letter is encoded depends on its position in the message. Figure 13.6.1 shows the basics of a keyword cypher – real-world implementations would use a longer key phrase and add extra layers of sophistication to how it is combined with the text, and can be very hard to crack *if you don't know the key*.

In Figure 13.6.1, the message is to meet in the hall at eight and the key word is 'wibble'. We assign numbers to letters, A=0, B=1, C=2... ignoring spaces, punctuation, lowercase, and so on. We start with the text letter M, letter 12, corresponding to the key letter W (letter 22). We add the key to the

Encoding

Text T	M	E	E	T	I	N	T	H	E	H	A	L	L	A	T	E	I	G	H	T
	12	4	4	19	8	13	19	7	4	7	0	11	11	0	19	4	8	6	7	19
Key K	W	I	B	B	L	E	W	I	B	B	L	E	W	I	B	B	L	E	W	I
	22	8	1	1	11	4	22	8	1	1	11	4	22	8	1	1	11	4	22	8
(T + K) mod 26	8	12	5	20	19	17	15	15	5	8	11	15	7	8	20	5	19	10	3	1
Cyphertext	I	M	F	U	T	R	P	P	F	I	L	P	H	I	U	F	T	K	D	B

Decoding

Cyphertext C	I	M	F	U	T	R	P	P	F	I	L	P	H	I	U	F	T	K	D	B
	8	12	5	20	19	17	15	15	5	8	11	15	7	8	20	5	19	10	3	1
Key K	W	I	B	B	L	E	W	I	B	B	L	E	W	I	B	B	L	E	W	I
	22	8	1	1	11	4	22	8	1	1	11	4	22	8	1	1	11	4	22	8
(C - K + 26) mod 26	12	4	4	19	8	13	19	7	4	7	0	11	11	0	19	4	8	6	7	19
Plain text	M	E	E	T	I	N	T	H	E	H	A	L	L	A	T	E	I	G	H	T

FIGURE 13.6.1 Encoding with a secret key.

Source: Spreadsheet available online [13G].

text using modulo (clock) arithmetic base 26, giving $12 + 22 \equiv 34 \equiv 26 + 8 \pmod{26}$. By adding the code for 'M' (12) to the code for 'W' (22) and working modulo 26 wraps around the alphabet to 'I' (8). To decode, we reverse the process: start with I (8), subtract W (22), add 26 to avoid negatives, with result $8 - 22 + 26 = 12$, the letter M.

The problem with this system is that, somehow, both the sender and recipient have to agree on the key and keep it secure. Imagine you want to send your credit card details to an online shop without them getting intercepted. You could encrypt them with a keyword, but then you have to send *that* to the shop without it being intercepted. This is the problem with 'symmetrical encryption' where the key to encode the message is the same as the key to *decode* it.

The solution is 'asymmetrical encryption' which uses a pair of keys: one to encode the message and one to decode, which allows what is known as 'public key encryption' (PKE). Using the credit card example, it works like this.

1 The shop sends you the *encryption* key (their public key) – but keeps the *decryption* key (their private key) secret.
2 You encode your credit card details using the public key, and send the result to the shop.
3 The shop decodes your information using their private key.

This way, it doesn't matter who gets to know the public key – it can literally be *published* – because all it is good for is writing encoded messages that can only be read by the holder of the matching, top secret, private key.

How can this work? Obviously, there must be a mathematical relationship between the two keys – so why can't an attacker work out the private key from the public one? The best-known method of public key

Pick two prime numbers p, q	$p = 5, q = 11$	
Calculate the 'modulus' $m = pq$	$m = 55$	
Calculate $f = (p-1)(q-1)$	$f = 40$	
Pick the 'public exponent' E_{pub}	$E_{pub} = 7$	This can be any number less than f that doesn't share any factors (apart from 1) with f
Public key (E_{pub}, m)	$(7,55)$	
Pick the 'private exponent' E_{priv}	$E_{priv} = 23$	Pick so that $(E_{priv} \times E_{pub}) - 1$ is exactly divisible by f
Private key (E_{priv}, m)	$(23,55)$	

Encoding with public key $E_{pub} = 7$, $m = 55$						
Message	W	I	B	B	L	E
t	22	8	1	1	11	4
$t^{E_{pub}} \bmod m$	33	2	1	1	11	49
Decoding with private key $E_{priv} = 23$, $m = 55$						
t	33	2	1	1	11	49
$t^{E_{priv}} \bmod m$	22	8	1	1	11	4
Message	W	I	B	B	L	E

FIGURE 13.6.2 Trivial illustration of public key encryption.

encryption – RSA – uses the properties of prime numbers. Figure 13.6.2 shows a (trivialised) example of how it works. The parts of the keys 7, 23, and 55 are all determined by the pair of secret prime numbers 5 and 11, but in order to calculate the private key (23,55) from the public key (7,55) you'd need to work out that 55 was the product of 5 and 11. This looks pretty simple (there aren't many prime numbers below 55 that it *could* be) but in real life, the prime numbers used are so large that the product is *hundreds* of digits long – and finding the prime factors of such large numbers is an enormous task even for modern computers.

Note that, apart from using trivially small prime numbers, there are many flaws in the scheme in Figure 13.6.2. For one thing, encoding text letter-by-letter like that with the same key makes the whole thing susceptible to the old high school letter frequency crack. To avoid this, chunks of the message would be represented as large numbers (imagine '220801011104' for 'WIBBLE' – although a base 2 encoding would be more likely) and each very large number chunk encoded in one go. Combine that with using far larger prime numbers and you begin to see the other issue with PKE. It uses a lot of computer power to do arithmetic on very long numbers. Even our trivial example resulted in numbers too large for Excel to handle. That's why Figure 13.6.2 example shows PKE being used to encode just the keyword WIBBLE from Figure 13.6.1. Real public key encryption is often just used to securely exchange a key to use with a less time-consuming cypher for the message.

Technical aside: You may have heard quantum encryption being discussed as something needed for the day when computers get powerful enough to crack public key encryption. In essence, the 'quantum' part provides a way of two people securely agreeing on a random number – in a way that is impervious to eavesdropping – which they then use as the key for a less exotic cypher to be transmitted by conventional means.

Having a good secret code is one thing, but how can you be sure who you are exchanging messages with? A useful feature of public key encryption is that it works the other way round – if you encode something using your private key, anybody can decode it using the public key. This can be used to verify the identity of someone you are talking to.

- Alice speaking publicly, announces "My public key is K_A".
- Bob meets someone claiming to be Alice. Bob gives this maybe-Alice a random word and asks them to "sign it" (i.e. encode it with their *private* key).
- Bob then checks the result by decoding it with Alice's widely known public key. If that works, it means that it must have been encoded with Alice's *private* key. This is as close to identifying maybe-Alice as the real Alice as mere mathematics can get. Of course, it all relies on the real Alice keeping their private key secret.
- This can easily be expanded into a two-way 'handshake' whereby Alice verified Bob's identity at the same time.

This is the basis of secure "digital signatures". If, say, you want to be sure that a website really belongs to your bank, this depends on some trusted third party whose job is to verify the identity of website operators and vouch for their public keys. If you want to create a website that shows up as 'secure' in people's web browsers, then you have to register with a "certificate provider" who must somehow confirm your identity. You then end up with a certificate containing your public key that has likewise been signed by the certificate provider to confirm that it is genuine.

Unfortunately, there is huge demand for these certificates and the process of issuing them is mostly automated – nobody will come around, check your birth certificate and meet your grandparents. In most cases all that will happen is that the certificate provider will confirm your email address and check that you are able to place a test file on your website. A scheme for 'extended validation' certificates where providers *did* do more stringent identity checks, and which displayed the name of the certificate holder in the web browser address bar has largely failed for various practical, economic, and political reasons – partly because it still made mistakes and also proved largely ineffective at preventing real-world fraud. So, when it comes to web pages, all the padlock icon really means is that your data is being encrypted, so it is hard for a third party to intercept. If someone has tricked you into visiting a fake website, you're on your own.

Digital signing using public key encryption has other valuable uses, though. It is a better way of implementing passwords: rather than sending your secret password to someone to be checked against their records, they have your public key and use it to send you a 'challenge' to sign with your private key. In systems like this, your password is often just an extra precaution to protect the file containing your private key and neither your private key nor the password ever need leave your possession.

13.7 How does this fit in a mathematical literacy curriculum?

The chapter is very much about mathematics that students should perhaps *know about* without needing to 'know how' to implement the details. Perhaps these topics are best seen as an attempt to promote *curiosity* (see Chapter 11) in the way computers actually solve problems and how subjects like sorting and organising data are actually an important part of modern mathematics. Even if some of the material in this chapter is too challenging for the typical classroom, wherever subjects like series expansions, logarithms, networks, permutations, or prime numbers *do* arise in the curriculum, it is well worth finding time for a brief discussion of their significance in computing.

Acknowledgements

Figure 13.3.2 is a screenshot from the freeCODAP tool © 2018 The Concord Consortium – the context comes from the Australian Academy of Science's *reSolve: Maths by Inquiry Project*. See link [13E] below. Figure 13.5.1 is from the XKCD comic strip by Randall Munroe [13F] and is available under the Creative Commons Attribution-NonCommercial 2.5 Licence.

References

Knuth, D. E. (1973). *The art of computer programming, volume 3: Searching and sorting*. Addison-Wesley.

Press, W. H., Flannery, B. P., Teukolsky, S. A., & Vetterling, W. T. (1986). *Numerical recipes: The art of scientific computing*. Cambridge University Press.

Links to useful material

To visit any of these links, scan this QR code or visit ltml.mathlit.org – append the link code to go directly to the entry – for example, ltml.mathlit.org/13A

The original source links are given below for attribution purposes:

[13A] *Simple Quadratic Equation Solution* (Spreadsheet & code)
https://ltml.mathlit.org/13A

[13B] *Rounding Errors* (Spreadsheet & code)
https://ltml.mathlit.org/13B

[13C] *A More Accurate Spreadsheet Quadratic Solution* (Spreadsheet & code)
https://ltml.mathlit.org/13C

[13D] *Taylor's series* (Spreadsheet)
https://ltml.mathlit.org/13D

[13E] *Monte Carlo Simulations* – reSolve & The Concord Consortium
https://resolve.edu.au/monte-carlo-simulations
https://codap.concord.org/releases/latest/static/dg/en/cert/index.html#shared=50813

[13F] *Password Strength* – xkcd.com
https://xkcd.com/936/

[13G] *Encoding with a secret key* (spreadsheet)
https://ltml.mathlit.org/13G

14

REFLECTIONS

Half a century ago one of us set out an agenda for bringing 'the real world' into school mathematics (Burkhardt, 1981). Then, together in Nottingham University's Shell Centre for Mathematical Education, we began to develop teaching materials to support this aim (see, for example, Burkhardt et al., 1980). Since then, each of us has made mathematical literacy an important part of our work through a series of research-based design and development projects [14A]. Some of the fruits of these projects have appeared in the various chapters of this book.

In writing this book, our aim has been to draw together into a coherent whole the insights from our contributions and those of others to the teaching of mathematical literacy. We have provided access to a diverse range of rich lessons that teachers have found to work well in getting their students to see mathematics as useful for their current and future lives. We have set these lessons in a framework for mathematics that is firmly context-focused, integrating the 'know how' and the 'know about' of mathematical modelling and data analysis within a critical thinking approach. Experience suggests that this helps students to develop a productive disposition towards looking at the world with a mathematician's lens.

We believe that teachers who work with this approach will find it as useful and enjoyable as others have. In this final chapter we reflect a little more deeply on some of the key issues that have come up in previous chapters.

14.1 On the context-focused mathematics framework

At the outset of writing this book, we saw that mathematics educators had devised multiple informative frameworks to organise the different components

DOI: 10.4324/9781003303503-15

of mathematical literacy (or numeracy, quantitative literacy, statistical or data literacy, depending on local terminology). Individual frameworks highlighted different components and linked them in various ways, but there is considerable agreement between them. They all included consideration of the context, mathematical and statistical knowledge (concepts, skills and strategies and usually use of tools), along with aspects of productive disposition (especially confidence) and underpinned by critical thinking about the context and the mathematics. These components seem to describe the task of teaching for mathematical literacy in schools well.

We wanted this book to show mathematical literacy in practice, illustrated by situations from real-world contexts that may be encountered by young adults. All components of mathematical literacy were evident within the set of examples that we explored. Yet this process highlighted for us the value of 'knowing about' as well as the school focus of 'knowing how'. In many cases a mathematically literate person (i.e. one who could make good decisions about a situation informed by quantitative thinking) needed a strong 'know how' of the basic mathematics/data literacy/tool use, but a more general 'know about' across more advanced ideas. By advanced, we mean mathematics beyond what is generally taught in the compulsory years of schooling up to about age 14. 'Know how' for advanced mathematics definitely contributes to the mathematical literacy of people working in quantitative professions, but for others, 'know about' is sufficient.

For this reason, we developed the context-focused mathematics framework (see Figure 1.1.1) for this book to include a wide ranging 'know how' component and a wide ranging 'know about' component. These two components each encompass knowledge, concepts, skills, and processes related to mathematics, statistics, data science, and digital tools. Some chapters in familiar areas – on planning and money, for example – have a 'know how' emphasis, with plenty of opportunity for modelling or calculating. Others, in more complex areas like climate change and inequality, are mainly 'know about', designed to support critical thinking about the data and the models that experts create, their assumptions and consequences. Life-related contexts have been the focus throughout.

This raises important questions. Teaching mathematics 'know how' for mathematical literacy is well researched and there is broad agreement on the special emphases required for mathematical literacy, outlined in Chapter 3 and evident in almost every chapter of the book. But how can we teach the 'know about' component well? Could the know about component be written into curriculum and could it be assessed? There are many questions. Does 'know about' accumulate from just a little 'know how' exposure (e.g. learn a little coding to understand how computers do what they are programmed to do). Perhaps 'know about' needs substantial 'know how' to see the fundamentals of a topic and remember them (e.g. to appreciate the importance of functional behaviour). Would including 'know about' goals in a teaching program (for example, by writing an essay on how weather is forecast or the dilemmas raised by inequality) make a lasting difference, or simply waste time. We know that spending a lot of time on routine practice (e.g. of arithmetic or algebraic skills) does not result in 'know about' (or, indeed, long-term 'know how') but is it possible to produce 'know about' without serious attention to some 'know how' skills? And the answers may be distinctly different for the many different things a mathematically literate person may need to know about. We have shown examples of how this has been done.

Another important decision in the book is to present theory-driven mathematical modelling and data-driven investigation as two approaches to answering a real-world concern. Some of the chapters lean to the theory-driven side, and others are data-driven, but all of the chapters show how intertwined these approaches can be in practice. A surprising realisation for us was to observe the great dependence of society today on *measures* that are, at heart, mathematical models of an abstract concept. These are used to compare or to track changes of things as diverse as the health of a national economy, how well washing machines perform, obesity, and fairness of society. Each depends on creating a mathematical model that identifies significant variables (and devising ways of measuring them) and then identifying the relationships between the set of variables that might give a useful measure. In this way, we see that a great deal of data is actually the product of mathematical modelling, further intertwining the theory-driven and data-driven approaches.

14.2 On the roles of technology

Technological advances have always been used to facilitate 'doing mathematics' – from making marks in the sand, via using an abacus to the latest computers – silicon to silicon in four millennia! New advances in these technologies have often raised concerns about the loss of familiar skills. In a speech on administrative reform Charles Dickens (1855) referred to the use of notched splints of elm called 'tallies' to keep the Exchequer accounts, which continued until 1826 even after an enquiry "by some revolutionary spirit" asking whether, "pens, ink, and paper, slates and pencils, being in existence, this obstinate adherence to an obsolete custom ought to be continued" – a situation not dissimilar to the resistance to the introduction of calculators in school mathematics classes a couple of centuries later, which has continued to the present day. In retrospect, each technological advance enabled a person to do more with less effort. Tables of logarithms were a prime example, enabling more precise calculation for navigation and many other purposes. In the last half century, we have stepped up from formula books to calculators of increasingly sophisticated types, and to computer software from GeoGebra to Mathematica and Wolfram Alpha. Each new tool has *shifted* the skills and knowledge required by someone to use them effectively to apply mathematics in understanding the real world. Recent 'large language models' (LLMs) like ChatGPT have the potential, used intelligently, to significantly increase the power of the individual as a mathematically literate person – although the initial concern from education seems to focus on their potential as a way of cheating on essays and dissertations.

But first on the basics. The school curriculum inevitably moves at a slower pace than the technology that is widely available to the population at a reasonable cost. But spreadsheets, for example, have been standard tools on home and work computers *for about 30 years*. As the chapters of this book illustrate, they are extremely useful to support a wide range of calculation and exploration. Making spreadsheet use a normal part of school mathematics is surely long overdue. There are, of course, other examples.

Looking forward, the current capabilities of LLMs could be indicating a major shift in those skills and knowledge may be required. If you can frame a problem in mathematical language (a formula, for example), then you can probably get a solution from a spreadsheet, or even ask Wolfram Alpha. But with LLMs, we may be able to ask open-ended questions in (non-mathematical) 'natural language' and get

a direction on how to understand the problem,
b mathematical solutions, and
c explanations on how to interpret and understand the answer.

For example, backpackers are using ChatGPT to plan their trips between one country and the next, giving travel arrangements within a budget, sites to see on route, and so on. People are using ChatGPT to help plan a party, including organising a schedule, managing guest invitations, budgeting for food/drink, and so on. It is used to get estimates of physical quantities: "estimate the force on the rope when a rock climber falls from 1 m above the bolt", "estimate how far away the horizon you can see is when you are on a boat surrounded by ocean". You can give ChatGPT a spreadsheet file and ask it to produce a basic data analysis.

These tools may make a difference to being a teacher – "Write me 10 questions on factorising quadratics"; "Find me a problem suitable for a one week assignment for year 9 involving similar triangles". Better check them, of course!

These new capabilities pose new potentially great opportunities for people to benefit from applying mathematics for the betterment of their life and their decisions, but they also surely pose new challenges and demand new skills of the kind we have foregrounded in this book. How do you interface effectively with LLM services – the next generation of the challenge of learning how to 'Google' effectively? How do you cross-check what an LLM tells you?

What ways are there to verify answers? Since LLMs sometimes give wrong answers, corroboration is essential. The list of new challenges is long.

Beyond this are the multiple varieties of potential damage that LLMs can do to society in plagiarism, malign messaging or just misinformation. This has already been exemplified by much less powerful and plausible social media. How this can be prevented – or, more likely, mitigated?

Critical thinking by mathematically and socially literate people surely has a role to play in the effective and positive use of LLMs – and is one thing that these systems *cannot* currently do for us. If students can easily 'cheat' by getting LLMs to complete their assignments, then the problem may be that those assignments are not requiring sufficient critical thought.

We end this forward look optimistically, with the last line of George Bernard Shaw's 'metabiological pentateuch' *Back to Methuselah* (Shaw, 1921).

> And for what may be beyond, the eyesight of Lilith is too short. It is enough that there is a beyond.

14.3 Developing professional expertise

It is clear that the variety of classroom activities and the range of contexts that we have described above will ask that many teachers of mathematics move beyond the comfort zone of their current practice. The sense-making approach, summarised in TRU and the context-focused mathematics framework, implies a broader range of concerns than is traditional in school mathematics – 'know about' as well as 'know how', with critical thinking about both the context and the mathematics and, above all, a focus on what the student is doing and the affordances the activities offer.

Some teachers welcome the opportunity to make mathematics more useful for their students; for them we have shown a wide range of both classroom and professional development activities that offer effective support for broadening the range of their expertise. Others are unenthusiastic – their job is demanding enough already. Yet there are cogent reasons for all teachers and teacher educators to explore this important domain with their students. For most students, mathematical activities focused on life-related contexts give added meaning to the mathematics, and motivation for developing their conceptual understanding and procedural skills – and they enjoy the activities. For the individual, there are benefits in their personal, work, and civic lives. For society the primary goal of school mathematics, and the justification for its large time allocation, is to produce appropriately mathematically literate adults for all levels of the workforce.

We hope we have encouraged all teachers of mathematics to move, however gradually, to include mathematical literacy as an aspect of many of their lessons and the focus of some.

And finally...

We have learned a lot in writing this book – and enjoyed the process. We hope that you, the reader who has got this far, have found this exploration with mathematics of some aspects of the world we live in enjoyable and informative – and that your students will too.

References

Burkhardt, H. (1981). *The real world and mathematics*. Blackie-Birkhauser.

Burkhardt, H., Treilibs, V., Stacey, K., & Swan, M. (1980). *Beginning to tackle real problems*. Shell Centre Publications.

Dickens, C. (1855). "Administrative Reform" (27 June 1855), Theatre Royal, Drury Lane Speeches literary and social by Charles Dickens (1870) pp. 133–134.

Shaw, B. (1921). *Back to Methuselah: A metabiological pentateuch* (Vol. 16). London Constable 1922.

Links to useful material

To visit any of these links, scan this QR code or visit ltml.mathlit.org – append the link code to go directly to the entry – for example, ltml.mathlit.org/14A

The original source links are given below for attribution purposes:

[14A] Past projects by the authors
https://ltml.mathlit.org/14A

INDEX